Mathieu Bourgois

About the Author

ALISON WEAVER received her MFA in creative nonfiction from the New School in 2004. She is the copublisher of the literary journal *H.O.W.*, the proceeds of which go to needy orphans worldwide. Her work has appeared in *Small Spiral Notebook*, *Opium* magazine, *Red China*, and the *Fifth Street Review*. Weaver lives in New York City.

GONE
TO THE
CRAZIES

A Memoir

ALISON WEAVER

HARPER ● PERENNIAL

NEW YORK ● LONDON ● TORONTO ● SYDNEY ● NEW DELHI ● AUCKLAND

HARPER ● PERENNIAL

A hardcover edition of this book was published in 2007 by HarperCollins Publishers.

FIRST HARPER PERENNIAL EDITION PUBLISHED 2008.

Designed by Kris Tobiassen

The Library of Congress has catalogued the hardcover edition as follows:

Weaver, Alison.
  Gone to the crazies : a memoir / Alison Weaver.—1st ed.
    p. cm.
  ISBN 978-0-06-118958-6
  1. Weaver, Alison. 2. Substance abuse—United States—Biography. 3. Recovering addicts—United States—Biography. 4. Problem youth—United States—Biography. 5. Problem youth—Rehabilitation—United States—Case studies. 6. Children of the rich—United States—Biography. I. Title.

HV5805.W43A3   2007
616.86092—dc22
[B]   2007061009

ISBN 978-0-06-137401-2 (pbk.)

08 09 10 11 12 WBC/RRD 10 9 8 7 6 5 4 3 2 1

To the memory of Jay Kistler,

I'm sorry that you're gone,
that you're not still somewhere,
your nicked hands chafing together for the next thrill,
your hair uncombed, wild blond.

I'm sorry that your guileless blue eyes are not still
blazing,
Your deep voice rumbling through the streets,
crawling down alleyways in the blue night,
echoing from city rooftops, inside water tower shadows
and fire escapes.

I'm sorry that you're not still speeding barefoot,
grass twisting your longest toe,
dirt stuck in neck creases, crashing through riptides,
sprinting over dunes,
a streak of black canine trailing.

The first Weaver Christmas card

CONTENTS

III. SUCH DESPERATE, DIRTY PEOPLE

PROLOGUE

The morning my parents send me away, the sky is a washed-out blue streaked with salmon. The air is tight and cold; morning frost blankets the ground. The dark green hedge that slithers around the property is dotted with beads of hardened water that glitter under the winter-white sun. Nothing appears out of the ordinary. I hear my eighty-year-old father shuffling through papers and mumbling to himself in the next room. In the bathroom, my mother is rushing. She sighs while dabbing liquid foundation on pink blotches that dot her face, removing curlers, sweeping her hair into the usual combed waves.

I suck down a cigarette on the balcony outside my room. Its orange and black ash crackles as I pull deeply, imagining the dirty smoke seeping through hundreds of clean, white fibers, expanding inside my lungs and absorbing into the tissue that encloses them. My stomach hurts. I feel like crying. I'd felt like crying from the moment my parents showed up on the campus of my boarding school and whisked me back to their house in Connecticut. I hadn't been home from Berkshire in months, but without any real explanation they had tossed clothes in a bag, muttering something about an appointment with an educational consultant, and directed me toward the station wagon parked outside my dorm. I didn't put up much of a fight. I'd had appointments with "educational consultants" before, and besides, we didn't have baths at boarding school. I thought I could take one if I went home.

"Ali, are you ready? We're running late," my mother calls.

I walk down the stairs and catch her eyes in the hallway mirror. She is now dressed and ready to go, her eyes murky and red like bloody toilet water, her lips pressed thin, pale, and dry. She lifts a bottle of Binaca to her mouth and sprays. She turns her chin from side to side, smoothing the collar over her cardigan. This is the mother I will always see when I think back to my adolescence: a woman in silk button-downs and black wool skirts, with the strongest peppermint breath I'd ever smell.

"Ali, please don't wear that," she says, frowning at my apparel: a black T-shirt, torn jeans. She lifts a cup to her lips. "It's inappropriate."

"Drinking awfully early, aren't you?" I say.

"It's Mr. and Mrs. T mix without the vodka, dear."

"Then let me have a sip." I reach for the cup.

"Don't be ridiculous," she says casually, turning it from my grasp. "Daddy is waiting."

The roads are clear, free of cars and fog, and I can see the sun rising from behind the distant hills. It looks like the sort of day that would make any person grateful for being alive. My mother drives in jerky starts and stops, and my father, head buried in a flowchart, sits next to her in the passenger seat. There is a faint murmur of something coming from the speakers, but no one cares enough to turn it off or make it audible.

When we arrive, the educational consultant is waiting. He is tall and robust with a tropical tan and a track-running body probably shaped at a state-of-the-art sports arena, like the one my father donated to Princeton years ago.

"Alison, why don't you come on into my office?" he says. He wears a finely tailored pinstriped suit and a navy silk tie with colorful trumpet-playing bears marching across it. He looks like every man who works for my father. "We can have a chat."

His office is large, with two black leather sofas and one wooden coffee table. Tulips sprout from a vase in the center of the table, sandwiched by two fans of magazines and school brochures. On a small circular table near an absent assistant's desk, a gurgling coffee maker disperses a nutty chocolate aroma.

I glance in the direction of my parents. They are holding hands. My parents never hold hands. Something in the air feels strange. I have an impulse to run. The Metro North Train Station is only two blocks away, but I have only six dollars and the ride to the city, I know, is eight-fifty.

"Alison, we both know why you're here," the educational consultant says once we are inside his small square box of an office, the door shut on my mother and father who remain in the waiting room. "Your parents are concerned about your future at Berkshire boarding school."

"My parents tend to overreact about things," I retort.

He doesn't like me. I can see it in his hard glare and the rigidity of his jaw line. I can see it in the way his hands are clasped, each tanned knuckle white with the suppressed rage he secretly feels for kids like me. I can't say that I blame him. I must look frightening in my torn, soiled jeans and "inappropriate" T-shirt, my black canvas school bag hanging low on my shoulders with its iron-on patches of teen bands splattered across the front, smacking gum, long, stringy bleached hair falling directionlessly from my head. I must look like a real mess to such a clean, wholesome man. Here among the groomed lawns and flourishing shrubberies of sub-urban Connecticut I stand out far worse than I ever did on the streets of Manhattan. I imagine he has a wife and two children and a cat and a dog waiting for him back at home, and I am sure he wants to get rid of me and my nastiness as quickly and efficiently as possible so he can get back to his Saturday morning routine.

"Your parents have asked me to help them find a school that will pro-vide the type of nurturing and discipline you seem to require," he says, his words smooth and practiced. "This is the school that we've decided would be best for you."

He hands me a brochure. A large cabin nestles in a bright green valley and a rainbow stretches from one end to the other on the front cover. Printed across this offensively bucolic picture is a name: *The Cascade School*. I leaf through a few photographs of teenagers crying while hugging or laughing with linked arms. Words of wisdom are scattered between pictures—*integrity, love, power, trust, kindness*. Below an image of a flourishing tree is a promise: *As a community we acknowledge the true potential of humanity and the nobility of the struggle toward a sane, caring, and enlightened world.* The words seem silly to me, and I force my eyes to slide across the glossy page. I lean back in the chair. A smirk or some other indication that I find the statement laughable appears across my face and the consultant grabs the book from my hands.

"It's a therapeutic rehabilitation program in the mountains of northern California," he says. There's something disturbingly triumphant in his tone. "You've already been enrolled, Alison."

It's only then that I see, through the half-shuttered window, a black limousine waiting in the driveway. It wasn't there before. It is shiny and dark, with a chrome grate slapped onto the front that makes it look threatening and alive.

"What do you mean I've already been enrolled?" I ask, unable to control the panic that slips into my words.

"Exactly what I said—you've already been enrolled."

"Well," I say, "you better unenroll me because I'm not going."

"Yes, you are, Alison," he replies. "It's all been arranged."

And as if on cue, two large men in flight jackets appear at the door. The educational consultant nods toward them, motioning in my direction. His hands have finally unclasped, the whiteness gone from the knuckles. The men step forward in unison, my parents behind them. And I notice that my father's face is wet. He is crying. I have never seen him cry. He looks haggard and old, his skin hanging jellylike from his face. The thin layer of gray hair covering his scalp is wet with perspiration, exposing years of spotty sun damage.

"Daddy, please don't send me away," I say instinctively. I get up from the chair slowly, trying to keep my composure. My throat is so dry I can hardly swallow. My cheeks are burning, stinging, my eyes filling with tears. Breathing becomes challenging; air is limited or something is holding it back. I can't believe this is really happening to me. This can't really be happening to me.

"I have to," he says. He won't meet my eyes. My mother is standing behind him with her hand on the doorknob. She is staring at us, and tears glint in her eyes, but the tears serve as a wall separating her from the scene as well. She watches and listens but lets none of it inside.

"Why?" I cling to the tweed of my father's arm. I can smell him—damp, wooly, and alcoholic. I haven't smelled him for years.

"Your mother and the doctors say it will help you. They know what's best."

It's all too much. The chilled air of the office perfumed with vanilla deodorizers, the tick of the clock radio on the consultant's desk, my father's tears, my mother's composure, the mundane noises of suburbia carrying on behind those dustless blinds all suck me dry of the small amount of self-control and sanity I still manage to maintain for occasions like this. They suck it right out of me like I imagine a mass of cells that is not yet life to be sucked from the uterus: brutal, relentless, gone. I begin to slap my face like a mad woman, right cheek, left cheek, moving slowly upward until my hands are fists pounding into the flesh that covers my skull. Fists believing that if they pound hard enough this will all end. Not just that moment in the doctor's office but everything before too: all the other doctor's offices I'd frequented over the years, the expulsion from school, the alcohol, the drugs, the emptiness; all of it.

Stop that, Ali. Stop it now! I hear my mother say. But then my hands are no longer enough, I need something harder, less malleable, so I move toward the wall, slamming my head into it again and again, swearing I will kill myself. *I will,* I'm screaming. I can hear myself faintly. *I'll kill myself if you do this. I'll kill myself if you send me away. How can you just send me away?*

Voices become distant and unrecognizable. I am underwater, cocooned; not the girl whose hands are fists of rage, whose mouth roars animal-like. Whose parents are staring at her as if she's a monster. People yell *let's move* and *go quick* and *don't bother comforting her.* Unfamiliar arms pull my hands behind my back and cuff my wrists in cold metal rings. My body reels and shrills as if tossed into the flames of Hell. I can feel hands, breath on my face, sweat, tears. I can see white walls, carpeted floors, wooden furniture legs, feet in rubber boots, bodies dressed in winter coats. Then I am tossed into the backseat of the limo I'd seen earlier. I am silenced by the slam of the car door but continue to scream, if only for myself.

Time passes. One hour, ten minutes—I can't say. Eventually, the door to the office opens and I watch as four adults move toward the car. They are talking, nodding, rubbing fingers on chins or across eyebrows. My mother's stance shifts from one foot to the other; she rubs the side of her hip, her lower spine. Her back must be aching. The big man who manhandled me into the limo towers over my parents, shaking their hands. The consultant slips his silk tie inside his suit, straightens his lapels, and subtly glances at his handsome gold watch. My mother's polished lips open, *Thank you,* while she nods her head in agreement to whatever he tells her. My father has regained his composure, and thus is no longer present. He's thinking about the garage roof that needs painting, the trees he needs to cut down, the celery soup and melba toast that he'll eat for lunch.

I am curled into a ball in the backseat as the car pulls out of the driveway, crying quietly now, yanking at the roots of my hair, digging my fingers deep into my flesh until my palms are crossed with lines of little red crescents. Time passes and I quiet, exhausted. We zoom down the Merritt Parkway, the exits getting smaller and smaller, Connecticut turning to New York, green turning to brown. I am sick of prettiness. I welcome the dirty city. It has become gray and humid with a light yet continuous drizzle. Outside the car, life goes on; dogs are walked, children taken on Saturday outings. Somewhere my parents serve lunch.

The flight to San Francisco is long. My giant escort intermittently pats my leg or hands me a tissue from his Kleenex travel pack. Ignoring his gestures, I wipe my nose on my sleeve and curl further into the ice-cold window. The clouds float beside me. I think the world seems peaceful from a distance, peaceful surrounded by the soft, calming whiteness of cloud. A stewardess in a maroon vest and navy blouse walks down the aisle stopping to check the overhead compartment above me. I close my eyes.

We land in Redding, California, at 4:07 PM. A medicinal white van waits for us in the airport parking lot. The driver of the van is an obese man in a tight, white T-shirt. His stomach protrudes significantly past his gold belt buckle; and his face, round and swollen, is covered with spidery purple veins that spread across his nose and ears as he and the escort stand exchanging pleasantries.

Redding looks like it's dying. A vacant town stretching over many miles. Dormant shops are thrown here and there, and peeling houses leak moldy couches and chairs into their front yards. Rusted pickups line the streets, and boys with mullets huddle in groups outside the closed Dairy Queen. A sign above the ice cream shop says: CLOSED FOR THE SEASIN. REASON, FREEZIN.

At the edge of town we turn onto a dirt road that twists through miles of thick wooded acreage. Fifty minutes later a log-cabined village appears: eight cabins shaped like perfect square boxes, each topped with a triangular metal roof. Names like Avalon, Gladrial, and Marie-Clare are written in oversized pastel colors above each door. Snow carpets the empty expanse surrounding the cabins, and a swirling concrete path matted with boot prints cuts through the blaze of white. In the distance is a bridge, a frozen pond, larger log cabins, and towering mountains, jagged like the edge of a saw or a torn piece of cardboard. Massive black rocks protrude through the distant ash-colored snow. Had I arrived there under different circumstances, I might have found comfort in the bareness of the land and the wildness of the mountains, but instead, the austere beauty adds to the horror of my situation. The driver makes a sharp turn through two steel gates, and to the right of the car I make out a rock with the words THE CASCADE SCHOOL etched in gold.

On the nearest cabin deck is a petite, mousey woman with twinkling blue eyes and silky brown hair that lands evenly below her padded shoulders. Poised and overly courteous, she bounces down the splintering steps, greets us, and shows us inside to a waiting room. The burly escort hands her some papers and leaves the building without another glance in my direction. I don't see him again. I watch her scan the papers—presumably information on me. Her silk blouse is neatly tucked into a long wool skirt, and as she flips through the pages, she absentmindedly fiddles with the skirt zipper that seems to be caught in some fabric.

"How was the flight?" she asks me.

"Fine," I say.

"I'm Janice Toffenheimer, head of admissions at Cascade. I met your parents a few months ago. Lovely people," she says.

"Yes," I agree numbly.

They'd been planning this for months.

A few minutes later, a short, stubby redhead bursts into the log cabin. Her hair is a profusion of perfect little ringlets, and she beams with delight at my presence, as if the sheer sight of my disheveled self somehow makes life worth living. Janice lists the events that are to take place over the next few hours. She says I will have to go to the "Welcome Center" and be strip-searched. She says they have to make sure I am not hiding drugs, knives, or weapons of any sort. She says my belongings need to be approved.

Two middle-aged women walk through the front door with linked arms and matching blue cardigan sweaters. They look at Janice and smile; Janice smiles back. Outside the window I see a group of children leaving a building, their heads drooping forward like dead tulips, their faces blotchy and red with grief. Some link arms; others walk alone. Three boys run between them throwing snowballs at one another and laughing; someone yells—*Quiet, no running!*—and in the distance, I hear sobbing, loud innumerable sobs. The sun slides behind the mountain peaks.

"You'll get used to the way we live here," the redhead says as she leads me outside. "You'll make new friends, and you'll even get a whole new family. It won't be so bad. I used to hate my life, hate myself, and now I embrace every day with such gratitude."

She walks along the snowy path buoyantly, with the demeanor of an officious saleswoman at a second-rate department store. Her name is Rona Crane.

"You know you sound nuts, right?" I ask her.

"No," she says. "Just enlightened. You don't get it yet."

"*Enlightened*," I repeat, incredulously.

"Yes," she says.

I dig a Marlboro Light from my pocket.

"Oh, no, you can't smoke here!"

The smell of sulfur briefly envelopes us as my match flares up. "I won't say anything if you don't," I say, taking a drag.

"I'm sorry, Alison. It's a rule, and it is my responsibility to enforce it," she says.

I smoke anyway. She flitters nervously beside me the whole time, blabbing on about integrity and values, her hands flapping in anxious little jerks. Days later I am still waiting for the repercussions, but apparently she never tells anyone. Maybe she was being generous, allowing me some slack on my first day.

"How d'ya do, ladies?" the woman behind the desk of the Welcome Center says as we enter the building.

"This is the new girl, Alison," Rona answers.

"Hey, little lady. Here, want a sucker?" She hands me a butterscotch candy. I haven't eaten all day, but I refuse on principle.

"No thanks," I say, passing it back.

"Alright then, hand me your bag," she says. Pleasantries dispensed with, she is brusque and businesslike.

She dumps my belongings out on the table in front of her. The smells of my old room fill the air: stale smoke, incense, coffee. She searches through pockets, glides her chunky finger across the hemlines of my pants, flattens socks, smells toothpaste and mouthwash. She works rapidly and efficiently like a factory worker on an assembly line, and within minutes she has confiscated most of my clothes and shoes, cigarettes, a glass pipe, and a plastic straw. She takes me into the center's bathroom, a thirty-foot-long oblong structure with a stone floor, crackling brown walls and an office chair with wheels between the sinks and toilets.

"Okay, take ya' clothes off, bend over facin' me, than turn and bend over with your back toward me," she instructs. "Walk back and forth, and after that, shower and dress."

Humiliation fills me. I take my shirt off, my bra. I drop them at my feet. A feeling of invasion rises, a familiar feeling of someone seeing something or touching something that isn't theirs, poking, scraping, ripping inside me. I begin to tremble. I pop the metal Levi's button from its hole, pull my zipper down, pull the jeans past my thighs and knees. I am not wearing underwear. Naked, I bend over as if searching for a contact lens. Blood rushes to my head. Shivering, I feel the cold mountain air inside me and the soles of my feet tingle and then slowly become numb. I choke back tears so as not to seem weak. I taste some vomit and swallow it down. I begin to pray—to what, I don't know. But I pray anyway. Pray for this to be over, pray for it to be a dream.

Bedtime is at nine o'clock. I lie curled into myself under a flimsy cotton sheet marked, in permanent black ink, *Property of the Cascade School*. I shake and quiver. My heart shudders in stops and stutters. I rub my frozen feet into one another, back and forth, back and forth, but they remain white and numb. Brutal mountain winds whip the walls of the cabin, howling, and I feel I am underground in a cold stone coffin floating somewhere between life and Hell. The rooms at Cascade are built like cheap coffins—thick, dark wood running along all sides, hard stone layering the floors, no heat, no fabric, no bath mats, just one long, hollow space.

I have never felt more alone in my entire life. Lightning gashes the tree outside our dorm and I hear the crash of a heavy branch falling onto the roof. The row of cloth dolls and stuffed animals on the shelf across from my bunk slouches into one another with missing eyes and worn, knotted hair. Their stitched-on smiles are frozen and frantic, as though they know something I don't and they seem to be trying to tell me about it. Their faces are calling to me: *run, little girl, run while you can*; and now they are screaming, standing now, jumping up and down. I lift my head to see if anyone else notices, but all I hear is the heavy breathing of other students, their snores, the subtle rasp of body against sheet. I must be hallucinating. I shut my eyes, pull the light sheet over my head and beg for sleep to rescue me, take me somewhere else; any nightmare is better than this one.

PART I

FURTHER TESTING NEEDED

Nobody ever overcomes the phantasms of his childhood. The man is the corrupt dream of the child, and since there is only decay and no time, what we call days and evenings are the false angels of our existence. There is nothing except sleep and the moon between the boy and the man; dogs dream and bay the moon, who is the mother of the unconscious.

—EDWARD DAHLBERG

MY MOTHER'S
HELPERS

Nurses raised me, one after the other until I outgrew them. I had my own room: It was large, with two tall windows and pillowed window seats upholstered in light blue chintz. On the walls hung framed Italian collages of monkeys and dogs in Georgian attire, and underfoot were plush cream carpets kept spotlessly clean. I grew up an only child despite my father's two previous marriages and earlier children—three daughters and two stepsons—who we saw once or twice a year.

I know my first nurse was named Wanna, though I remember very little of her. In the pictures that linger on the pages of photo albums, she is either pushing my steel-bodied British Silver Cross baby carriage, or stands as a figure in the distance ready with a bottle or a squeaky toy. In every shot she wears a white uniform, a navy cardigan, and pink fifties-style glasses.

After Wanna, there was Ms. Bee, who sued my parents, claiming that my rambunctious behavior in the bathtub put her back out. I don't remember her at all. Then came Isabel, who let me play on the electronic toys at Hammacher Schlemmer and crawl back and forth inside the thirty-foot configurable play maze, while she "fraternized" with the male employees. Isabel liked to hit. She'd hit if I didn't eat my Jell-O or wash my face in the bath. Once, as we played Battleship, she caught me peeking at her board and decided to give me the false satisfaction of sinking

With Dad and one of my many childhood nurses

all her ships before smacking me across the cheek. Isabel stayed until I was five.

But then Ilse arrived on a Friday morning in October. Mother always wanted her help to start on weekends so they'd be comfortable with my father's presence in the house. Mother also made it a point to hire older women who lacked any beauty or sex appeal. *Lead him not into temptation*, she must have thought, sipping from her wine glass.

Ilse was in her fifties. Her black hair was streaked with lights of silver and her eyes were a deep liquid brown, almost tragic in their intensity. Chubby, flushed cheeks, curved with sensitivity and exhaustion, held a perpetual smile in place, and her stocky, willful figure—unathletic yet surprisingly mobile—was stiffened by an arthritic ache on winter nights. At first, I was terrified of her touch. I froze when she kissed me goodnight and squirmed when she hugged me or held my hand. If she invited me into her room at night, I'd sit on the floor with my knees clasped to my chest and my back against the wall.

Ilse smelled of fresh sheets and thick, white Nivea cream, and the smell gradually instilled an unusual comfort in me. Eventually, we spent our evenings on her twin bed watching *Family Feud* or *The Price Is Right*. She introduced me to Chuckles candies and HoneyBee suckers with buttery insides, and taught me how to speak German. If I had a nightmare, she'd tickle my back until I fell asleep, and if her arthritic hands ached, I'd massage them with the thick, white lotion.

From the start, I lived most of my days in worlds that didn't really exist. The real world frightened me. I didn't fit. Other children were different: happy, approved of.

I chose to insulate myself from the outside by creating my own fantasy worlds, places where I felt alive and strong. I created imaginary friends. Some dwelled in the porcelain bodies of my dolls and others simply drifted about the playroom in ethereal form, handing me a piece of chalk when I needed one or being my opponent in a game of Chinese checkers. They played with me for hours in the attic, where I taught them spelling and math, chalking on the blackboard as they sat lethargically in a semicircle. I created entire block countries and governments to run them, relegating certain Cabbage Patch dolls to positions of power and giving myself the position of dictator. I directed productions of *The Sound of Music* every Saturday at 2 PM and sold handmade tickets. Occasionally, the cook or the maid would humor me and purchase one for five cents, but they showed up only rarely. The shows, however, went on anyway.

I was an adventurous child, outspoken and precocious. I didn't like fancy dresses and I didn't care for shoes, and I was constantly removing both whenever an opportunity arose. I liked dirt and making mud pies and wading through stagnant water to catch frogs and tadpoles. Blood didn't frighten me; neither did ghosts or dead people.

I could sit cross-legged for hours on the mossy stone wall in the middle of the strand of Connecticut woods that divided our property from the neighbors', surrounded by tall sapping pine trees and gargantuan oaks. Their roots seemed to tunnel so deeply into the ground, I imagined they came out on the other side of the earth. Sometimes, I'd pretend to be Huck Finn or Laura Ingalls Wilder or one of the fairies from *The Green Fairy* book. I was myself only in the briefest of moments, only when the outside world demanded it of me.

2.

UNREADINESS AND INAPPROPRIATE BEHAVIOR

When I was five years old, I told my first psychiatrist a fairytale about a girl named Princess Leia who wanted to kill a very bad man named Darth Vader and and sleep in a beautiful bed with many pillows and not have to live on Earth. I told her that Darth Vader had killed Leia many times already and that she was very afraid of him, and that the only way she could return to outer space was if Darth Vader were dead. She asked me why Darth Vader wanted to kill Leia, and I told her it was because Leia's mother said Leia was a very bad girl and when she tried to catch her and lock her up, she couldn't because she was very fast and ran away, so she sent Darth Vader after her.

When she asked me why Leia was a bad girl, I responded with genuine curiosity, saying, "Yes, why?" And I explained to her that Princess Leia didn't *mean* to be a bad girl, but that she couldn't help her behavior. I explained to her that Princess Leia was born in outer space and wasn't meant to behave like the people on Earth.

This first experience with psychoanalysis occurred after my second semester of kindergarten at the Spence School, an elite private school on the Upper East Side of Manhattan. My teachers had begun to complain that I was demonstrating "unreadiness and inappropriate behavior." I was

With friends at the Lyford Cay Club, Nassau, Bahamas

having problems with some of the other children, being too aggressive, pulling hair on the playground, laughing during lessons, and talking out of turn. They recommended I be sent to a psychiatrist, and my parents agreed.

My first doctor wore tailored suits and minuscule pearl earrings and slipped her stockinged foot in and out of a black pump during our sessions. Sometimes, when she thought I wasn't looking, she'd rub it against the leg of her desk to alleviate an itch. During our first session, I slouched in the velvet armchair, averting my eyes from her scrutinizing stare and wondering what we were doing there.

I knew that I was being tested on something, but I couldn't grasp what. Without the usual matching colors and shapes, addable numbers or spellable words that a child of my age ran into almost every day, I was lost. But I wanted to please her, to convince her that I was smart. I did exactly as she asked that day. I pretended to play with dolls. I did puzzles. I drew pictures. But she was unwilling to give me any indication as to how I was doing. At one point she pulled out a drawer and slid a yellow legal pad discreetly from it, removed a pencil from behind her ear, and scribbled some notes.

Whatever I did or said during that session led the doctor to believe that I suffered from "conduct and anxiety disorder," that I was unable to suppress my "bad behavior" or "running about," and that I lived "in a fantasy world that engenders in her nearly uncontrollable amounts of anxiety." At the conclusion of the doctor's notes from our first meeting, the pivotal meeting that was to place me inside or outside the confines of therapy, she wrote: *This material indicates that Alison feels unlovable, unwanted, bad, angry, and alone. I would say in conclusion that this child is very taken by her fantasies. These fantasies are created to escape the pressure of the outside world. She almost lives in these worlds that she creates. They are well organized and compelling but they arouse a large amount of anxiety too. She is currently very uncomfortable. Further testing is needed.*

When I came out of the doctor's office after that first appointment, my cold and beautiful mother waited for me on the edge of the long leather couch, flipping through a magazine, legs crossed, a Hermes scarf wrapping her neck.

"Lovey Buffins," she crooned, opening her arms and pulling me close to her, fiddling with my braids that had come undone, flattening my collar over my sweater and engulfing me in her familiar smell of coriander perfume, white wine, and Binaca. "You're a mess," she said.

Then she smiled as if to reassure me, but I recognized the smile as something else. Maybe it was the way her head shifted or the sides of her mouth rose, maybe it was the quaver in her lip, or her eyes that throughout my childhood always seemed to be glossy with the remnants of tears. The smile seemed deceitful, doled out solely for the purpose of comfort, employed only during moments when the reality of the situation was too distressing to discuss.

A feeling of shame blanketed me that night as I lay in bed, shame for whatever it was that brought me to that doctor. My stuffed bear lay tucked under the covers beside me, his floppy worn arms folded neatly over the blankets. The bear had a pocket in his chest where a bright red wool heart was tucked, and sometimes I'd take the heart out and pray to it, but I remember that night things seemed too hopeless to bother. Ilse's voice came through the beige intercom on the wall: "Go to sleep, Alison. You'll be tired for school."

I must have been talking out loud.

"I can't," I said.

"Try," she said.

And then I realized I was crying. My nightgown was soaked in sweat and tears, and my chest was so tight it hurt to breathe. I was terrified I'd have to spend the rest of my life hiding whatever it was that was wrong with me; the black monster inside my belly.

This night still sits on the periphery of my memory as clearly as it did that day in 1982. The soft night-light glow that seeped through the cracked bathroom door, the trail of condensed water flowing from the windowsill where the humidifier sat, the glow-in-the-dark stars I'd secretly pasted on the left corner of my ceiling, the taxi lights that intermittently passed by illuminating the room with a flash of yellow.

My mother appeared at the door.

"What's wrong with me?" I sobbed. "Why did you send me to that doctor? I don't understand."

"Nothing is wrong with you, dear," she promised. "The doctor is just going to help you behave better in school. Spence wants you to see the doctor."

I didn't believe her. I could see terror in her eyes. I knew I was scaring her, but I couldn't help it. As I screamed and cried, the color in her signature rosy cheeks faded, and she turned white with fear, and, for a moment, she just sat there looking at me. Her hair was pulled back with two silk-covered combs that formed a graceful wave around both ears and broke delicately at her shoulders. Her hands were clasped in her lap, long and elegant like an antique porcelain doll's.

"All right, all right dear. Calm down now," she said. "Daddy is going to wonder what the fuss is about. You don't have to go back, but let's not upset Daddy. Please try and go to sleep now," she tucked me under the covers and folded the top sheet under my neck.

"Ilse," she spoke into the air.

"Yes, Mrs.," Ilse's voice came through the intercom.

"Can you come and sit with Alison until she falls asleep? She's a bit worked up."

"Yes, Mrs. I'll be right in."

Maybe that was the night when her usual glass of wine turned from a healthy ritual of enjoyment to an escape. Maybe it was the first night that she finished an entire bottle of white wine while she watched *The Charlie Rose Show*. Or maybe she was already drinking vodka back then.

I lay awake for hours, long after she had left the room. Ilse sat on the window seat staring into the night, out across the East River to the flickering lights of Roosevelt Island or into the brightly lit windows of the oversized townhouse across the street. At times she paused, yawning, massaging her arthritic hands or glancing over to see if I had fallen asleep.

"Ilse," I whispered.

"Yes, Alison."

"Can you tell me about the Nazis again?"

"Not now, sweetheart. It's very late," she whispered. "Should I rub your back?"

"Will you tell me tomorrow?" I asked.

"Tomorrow," she said, climbing down from the window seat.

"*Gottin morgin*," I said.

"That's 'good morning,'" she said, smiling.

"Good night," I whispered.

"Good night," she said back.

A week later, my mother claimed to have forgotten her promise, and I was carted off to the doctor in tears. We never spoke about it again, and I spent one afternoon a week in therapy for the next four years.

3

MISS MANNERS

As a child, I was unable to recognize the strange social codes that my parents lived by because I existed inside of them, but now they glare up at me wherever I go. I watch the rich. I watch them play the game. I watch them trot out their fake smiles. I watch them when I visit my mother in her uptown apartment. I watch them walk down Fifth Avenue past spotless windows full of clothing pressed in perfect folds, gathered emptily around sculpted mannequins. I watch them climb the silver ladder, step by step, as pieces of themselves break away from their limbs, as they become nothing more than expensive clothes hanging off bodies without souls, trying to be loud and smell sweetly and shine brightly, but never amounting to anything at all.

I recognized some version of this when I was eleven years old.

We were going to a Carter Christmas party, a yearly event my family had been attending since I'd started at the Spence School at five years old. I was sitting on the jump seat of an Esquire limousine, staring out the window, watching as the doormanned buildings decorated with yellow lights and tinseled wreaths slowly crept by. The holiday traffic was thick and stationary. We passed precious, overpriced stores—stores that sold delicate linens, silk baby clothes, antiques, gourmet foods, and rare orchids.

I could feel rebellion brewing inside me. I hated the fancy dress I had on. The large, scratchy lace collar felt like an enormous doily around my neck; the sleeves were too short; the cuffs too tight. The red barrettes in my hair itched.

The Carter family Christmas party was held in an enormous apartment building on Ninety-fourth Street and Fifth Avenue. Rooms were sectioned off with different activities for the children; photographers snapped and crab puffs were passed on silver trays. Laura and Carrie Carter would mingle with the grown-ups, dazzling them with their refined Emily Post manners and exceptional vocabularies. They were intelligent girls, primped and pressed, obedient to their parents' every wish. They were the kind of girls my mother wanted me to be, and I noticed her covetous glare as she spoke to them, her mind working overtime, trying desperately to figure out what she needed to do to make me like them.

"What have you been up to, dear?" she asked Laura.

"Mostly school and ballet. I'm going to be in the New York City Ballet's production of *Swan Lake*," Laura replied, sipping her Martinelli's sparkling cider.

"Oh, how wonderful. Ali never really stuck with ballet." My mother turned to me. "Why was that, dear?"

"Because I was horrible at it," I said.

"Oh, stop," she said. "She never practiced. That was the problem."

"I did practice, Mom," I said. "I just wasn't any good."

My mother sighed, not knowing what to do with my impudence. She lifted a glass to her lips and drank deeply. I watched as Laura and Carrie Carter fawned over each other, delighting in the praise that had just been bestowed upon them, until I couldn't take it anymore and went to find the dessert table.

The nothingness of it all hit me as I stood alone in the corner of the Carters' antique living room with a handful of jelly beans in the palm of my hand, watching the grown-ups mingle, clicking big glasses filled with brightly colored liquid and kissing cheeks and talking grown-up talk. Chatting about their most recent vacations: safaris in Africa, rented yachts and European tours, or the latest triumphs of their soon-to-be lawyer, banker, or stock analyst. Mirrors on either side of the wall multiplied them, their fast-moving, lipsticked mouths, their darting painted eyes, the nips and tucks of chin and forehead. Loud, cheerful

At one of the many holiday parties we attended

music vibrated from the speakers. And it felt like a big spectacle to me, a huge, over-done circus performance. But then, as if I were looking at a painting covered in too much varnish, the top layer began to peel away, and in a flash I saw the dark and frightening emptiness that lay below the color.

By the end of the night I was so full and nauseated I could barely stand up straight. One of the Carters' nurses obstructed me on my hundredth lap around the candy table, wiped my mouth with a wet paper towel, dusted crumbs off my front, and hustled me out to the foyer where the adults stood kissing drunkenly goodbye. I nestled up to my father's side. He patted my head.

"C'mon, Rose," he commanded. "Let's go. Al's tired."

"Ring the elevator. I'll be right there," my mother replied.

My father and I walked into the elevator hall where striped English umbrellas shot from chic metal bins on either side. He lifted me to press the button, tickled me a little, and plopped me on the bench between the umbrellas. I laughed and tickled back. My mother's crane-like neck peeked between the heads of our hosts and noticed us.

"Don't get her worked up, Will," she called. "We'll never get her to sleep." Her voice was high and breathy and nervous.

My father stopped, rolled his eyes, and let me climb onto his back. I was almost too tall for piggyback rides—my gangly legs bounced against his knees and my lanky arms wrapped like a scarf around his neck—but still, I was thin and light, and at seventy-six my father was quite fit. Mother kissed the Carters on both cheeks one more time and we left. In the limousine, I dozed in and out of sleep against my father's arm as white tufts of cloud scudded across the gray New York sky. My parents talked in slurred sentences about the evening.

"Some of those parents are just awful. I can't understand why Spence is letting families like that into the school," my mother said.

"When did you become such a snob?" my father said, yawning, one foot extended forward.

"I'm not a snob, Will," she said. "I am talking about *quality* people."

"Oh, *quality* people," he said. "Since when did you become the authority on quality?"

"Stop being argumentative. You know what I mean."

"No, I don't. I like the Golds and the Burgers."

"They're perfectly nice. I didn't say they were not nice, but they're just not our style. They're too showy, too new money. It's just not the way respectable people conduct themselves in the world, with all that flaunting."

My father laughed. "Aren't you a regular Miss Manners."

Only one picture exists of my father as a boy. In it he is wearing a sailor shirt, white shorts that stop mid-calf, little white socks bunched at the ankles, and a cap with a navy ribbon bowed above his ear. When I was a little girl, I had a recurring dream of the boy in the picture and I, in a matching sailor dress, running though the woods.

I know very little about my father's past, and what I do know has come to me in snippets over the years, as flashes of information plucked from the echoing phrase of an aunt's sideways comment to my mother, or deciphered from raised eyebrows and half-overheard conversations between cousins or household help. I have spent years catching these fragments and placing them alongside the others, in hopes of making some sense out of his life.

What I do know is that he was born in 1912 in a two-family house in Jenkintown, Pennsylvania. I never knew my grandparents, and when I asked about them, all he could recollect was that my grandfather was a fisherman and a golfer, and that my grandmother took shooting lessons from Buffalo Bill's girlfriend, Annie Oakley, and played the banjo. He claimed we were related to the naval hero John Paul Jones and that one of my ancestors was the product of a love affair between Lady Hamilton and the Earl of Aran, but as with many of my father's stories, the line between fact and fiction is blurred.

In pictures of him as a young man he is tall and broad-shouldered with dark wavy hair, no hips, and stick-thin legs. He was the first of the Weavers to attend college, graduating *magna cum laude* from Princeton in 1934. He remembers that very clearly.

Married to wife number one shortly after graduation, he had two little girls by the time he was twenty-eight. World War II began in 1941 and he served as the youngest colonel on General Omar Bradley's staff. When

Dad, then and now

the war ended, he was requested to take a position with the Office of Strategic Services, later the CIA, in Washington, D.C. My father never talks about those years, but it seems he abandoned wife number one and his children in their suburban home in Blue Bell, Pennsylvania, justifying the action as a career move he couldn't afford to turn down. A divorce was granted one year later.

His time in the OSS was kept even more covert than the rest of his life, though I do know he roomed with J. Robert Oppenheimer. I tried many times to ask him about those years but he always said the same thing: *Al, if I tell you, I'm afraid I'll have to kill you.*

Six years later, it was 1960 and he was living in New York City working as an investment banker for Alex. Brown & Sons, married to wife number two, a popular fashion model. This marriage produced one child, a girl, and lasted until 1971 when he met my mother.

What knowledge I have about my mother's past is equally vague. Even now when I question her about her childhood she looks at me with an expression of bafflement, as if she doubts the accuracy of her own memories. To this day, she has no idea what her own mother died of—*some cancer I think. She didn't tell us. I don't think she wanted us to know.* Though my mother never speaks about her own mother and I remember very little of her, I have picked up oblique references toward volatile alcoholism and I don't believe she has any fond memories of this woman.

My mother was born into a middle-class family from Scranton, Pennsylvania. Her real father abandoned them when she was a child and her mother remarried two years later—though, in keeping with my mother's usual reticence, this "real" father had never been mentioned until one evening a few years ago, when by some shrewd detective work of my own I uncovered the fact that the man she'd previously referred to as her father was actually her stepfather. When I pointed out this obvious omission on her part, she stared blankly up at me, saying, *Didn't I ever mention that, dear? It must have slipped my mind. Why are you so concerned anyway?*

At twenty-two, my mother moved to New York City with the usual unjaded delight and optimism twentysomethings come to big cities with, and this is where she begins to remember her past in meticulous, if not exhaustive, detail. You see, my mother always had a taste for money and status, and as soon as it came her way, she began to follow it. Living on the Upper East Side, she claims to have attended parties thrown by George Plimpton, the literary icon of her generation. My mother couldn't tell you the difference between Joyce and Faulkner and if you told her it was Virginia Woolf who stuck her head in the oven and Sylvia Plath who filled her pockets with stones and walked into the river, she'd believe you. But she was pretty and charming and knew how to drop the right names and ask the right questions. She didn't date intellectuals; she dated neurosurgeons or bankers who smoked Gauloises and drank Glenlivet and dined and danced with her at private clubs that served raspberry blinis and ginger crème brûlée.

She met my father in 1971. They were introduced by a mutual friend who was visiting from Beverly Hills and insisted they both come to dinner at the Waldorf Towers. As the story goes, they fell madly in love and wanted to marry immediately, but my father was still technically married to wife number two. Three months after their meeting he purchased a 15,000-acre ranch in Smith, Nevada, where the divorce

Mom and Dad on the Lazy W Ranch in Smith Valley, Nevada

laws permitted him to quickly get rid of his second wife. He named the ranch Lazy W. A divorce was granted seven weeks later, and my mother and father were married in Reno, Nevada. He was sixty and she was thirty-two.

In the first volume of my father's self-published autobiographical trilogy, he includes a picture of my mother leaning against a tree in a beige V-neck sweater and white collared shirt, looking dreamy. It's captioned *Thank you, Mr. Beverly Hills!* in an offbeat thanks to the friend who introduced them.

The new Mr. and Mrs. William Weaver Jr. purchased a twelve-room apartment on the sixth floor of 25 Sutton Place in June of 1976. I was born one year later, and there is some argument as to whether I was a "lucky" accident or a mistake.

If you asked my mother now what kind of child I was, she would smile and say, *Ali was very easy. People always said—you have such a happy child, Rosemary. You're so lucky.* If what she says is true, then the memories of my first five years—and all that came after—seem suspiciously erroneous. The thing is, childhood memory is subjective and imprecise, composed partially of invention and interpretation. Sometimes, in fact, I am not sure if what I am recalling is my own recollection or a story someone habitually reminded me of that through familiarity became my own. Memory is a tricky animal, especially within my family where truth, reality, and myth coalesce into some indistinguishable miasma, but that is the fallacy of the human condition—that to understand who I am today, I must rely on what I only *think* I know of yesterday.

4.

MOM WEARS
TOILET WATER

I was sometimes happy, especially at the beginning, but the moments were brief and fleeting, never constant. When I shut my eyes and try to remember the good times, my weekends in Connecticut always come into focus. Our Friday afternoon drives were probably the only chunk of time I spent with both my parents at once. In the early years, it felt good inside that car. They were playful with each other, my father saying things to shock my mother as he tenderly massaged the nape of her neck, and she reprimanding him blithely as her lips broke into a smile. The air in the car always seemed light and buoyant, and talk flowed freely among the three of us as we passed bags of pretzels and cashews back and forth.

I'd stretch my lanky body out across the backseat, imprinting the window with my warm bare feet and watching the prints disappear, appear, disappear, appear, while singing rounds of "Row, Row, Row Your Boat" with my father. When we drove past the graveyard a mile from our property, he'd point to it and quip—*People are dying to get in there.*

Once we arrived, my mother would go into the library, where she would lie on the navy wool couch with a glass of wine, slicing open mail with a long silver letter opener and flipping through *W* or *House and Garden.* My father would go upstairs to his dressing room, change into work clothes, and head outside to garden. I would run to the attic playroom.

Of course, I didn't see much of my parents on these weekends. Ilse was always my companion for adventures. During the day, we would walk deep into the woods surrounding our property, as I tried my best to get us lost. Ilse knew the woods inside out, but she always pretended to be terrified that we'd never make it home again.

During the fall months, she'd rake gigantic piles of leaves together, and I'd wrap my body up into a cannonball and leap from a low tree branch into the thick pile. Sometimes she'd take me on rides in my red Radio Flyer wagon, pulling me along the long and winding gravel driveway, down to the greenhouse, where we'd check on all the talking plants, or over to the rock garden, where an entire village of dwarves lived.

Ilse indulged all the whimsical eccentricities that I now realize were necessary for my survival as a child. With each imaginary world that I was a part of inside my head, I was somehow safer, somehow more insulated from reality, where I constantly felt bad or wrong or misunderstood.

I believed then that a wild monster slept in my warm underbelly. A large, frightening thing that forced me to pull hair and talk out of turn and stick my fingers into cans of paint during art class. Many people hated me because of my monster, but Ilse saw it and loved me anyway. She loved me even after the hysterical tantrums and lies, even after the

With Ilse

stains on my dresses and the rips in my pants. On some level I believe she recognized where my pain was coming from and understood it—though she would never have acknowledged it out loud. This world was feeding her and clothing her and buying her Christmas presents and paying her doctors' bills, but she knew there was something empty and sad about it, and she knew that's why I'd reacted so badly. She knew I wasn't wired like the others and that pretending I was made me crazy.

Twice a month Ilse spent the weekend with her family in Middle Village, Queens, and on those weekends my father allowed me to trail behind him as he moved from one outdoor project to the next. Mornings were spent eating hot oatmeal in front of the dining room fire before my mother awoke, which was rarely before 10 AM. We'd hurry out into the foggy Connecticut dawn in work boots—his old and caked with mud, mine new with clean yellow laces. I remember my father best in Connecticut, behind the wheel of the John Deere tractor in worn red corduroy pants, always held up by a tattered polka-dot tie, and his bulky cable-knit sweater torn apart by summer moths but still wearable.

As a child, it seemed that my father was many men: the English cricket player in the green, yellow, and red striped cricket suit; the Scottish bagpipe player in kilt, sporran, and dagger; Fred Astaire in tails and a top hat tapping his way down the stairs and serenading my mother with *You're the top, you're the tower of Pisa*. In New York, he was always the businessman: the director of Allen Group Inc., IU International, and Gotaas-Larsen Shipping Corporation, partner of the investment banking firm Alex. Brown & Sons, and the president of Howe-Met. The man in impeccably tailored Saville Row suits. The man sitting in his corner office of the Equitable Building on 1260 Sixth Avenue sucking the edge of a piece of yellow loose-leaf paper or nibbling a number two pencil while adding numbers in his head. The man who lunched at the 21 Club, Le Cirque, or Judson's Grill and knew all the waitresses by name. The man who ordered twelve Malpeque oysters for lunch every day and drizzled them in minuet sauce. The man who loved oeufs à la neige and was affable to anyone he encountered yet subtly pragmatic in his dealings, always thinking and watching for the next opportunity to present itself.

In Nevada, he was the rancher. The man who, after purchasing his first ranch, became obsessed with farming and cattle and manically bought ranch after ranch across California and Nevada and Montana. There, he dressed in cowboy boots and snap-button shirts, and acquired a fondness for the scent of manure.

In 1989, when summing up how he arrived at his affluent position for the second volume of his self-published autobiographical trilogy, he listed the following: *bribes, charm, desire, education, guidance, hunger, imperviousness, intuitiveness, luck, milieu, intelligence, politics, scotch whiskey,* and *sex.*

But to me, on those early mornings in Connecticut, he seemed to float above all these titles and exist simply as my father. His intellect and exuberance fed a hunger for knowledge and life that I so longed for as a young child. Sometimes when I was with him, I felt he knew something no one else in the world did. He doled his precious secret out to me in little bits over the years, sometimes with a story or a grin of satisfaction for what he believed I understood, sometimes when we connected over something humorous that did not amuse the others.

Making my father laugh was one of the most rewarding activities of my childhood. There was something about his smile that, when sent your way, made you feel special and important. He had this guilty smile that was especially gratifying to receive because you knew he was smiling despite the knowledge that it was wrong. He'd push his tongue into his upper lip, bulging it slightly below his nose in an attempt to contain the laughter while leaning back into his chair in his best effort to look inconspicuous. In those moments, I felt secure in the certainty that there was at least one person in the world who loved me.

During the summer, my father and I would spend hours each day on the tennis court practicing lobs and volleys, backhands and forehands. He was an excellent tennis player and he wanted me to be one, too. He hired a private instructor to come to our house on Saturdays to give me lessons. The instructor gave me cans of Sprite and Snickers candy bars if I scored. My father would sometimes watch from his dressing room window while he was getting into his tennis whites for his own doubles

Dad and me at one of our summer tennis clinics

match later that morning, and every few shots he'd yell, *Harder Alison* or *Jump for it* or *Hustle up, hustle up.*

Sometimes, we took late afternoon drives to the library or the hardware store. Cows grazed near the winding country roads of Easton, and my father in his antique Mercedes, looking woodsy yet dignified, stared in wonderment at the mud-covered bodies, perhaps comparing them to his own herd in Nevada. We zoomed along Sport Hill Road at a speed we never reached with my mother in the car, the cold air from the open windows beating our faces with life, and I'd be filled with the giddy ecstasy of an outlawed freedom.

"Al, do you know what my favorite scent is?" he once said.

"No, what?" I asked

"Cow manure," he said. "Smells like money being made. I'd like to bottle it and sell it as a cologne."

"Well, why not?" I said. "Mom wears toilet water."

He burst into laughter at my confusion, and I sat in the passenger seat smiling a 200-watt grin, proud that I'd said something that made my father laugh so hard. He told that story for years to come, to anyone who would listen. The unfeigned veracity of my remark—my innocence, my lack of artifice—was precisely what he adored about me, and what he clung to until there was no more to be had. Even during the most horrific years of our family's dysfunction, he'd tell the story at various dinners, perhaps in hopes of reminding me who I once was. And though I was then both disgusted and deeply wounded by him, hearing him speak the punchline, his words half-tossed between the ice cubes in his scotch, gave me a brief feeling of contentment. His voice, though older and perhaps cracking slightly, was still authentically his, and somehow it rendered the weight of everything else between us remarkably insignificant.

5.

GONE TO
THE CRAZIES

I took my first photograph at eleven years old. It was of a homeless man asleep on a bench outside Central Park. His mouth was slightly open, revealing the jagged tips of four grimy yellow teeth, shiny with the gloss of plaque and various age-old buildup, and small uneven gaps that led into the dark rankness of his mouth. A brown canker sore pussed on his lower lip. He was wrapped in a thick, tattered beige blanket spotted with dark stains that reminded me of the inkblot shapes on the Rorschach test.

I climbed the wall behind the bench, tossed my backpack onto the ground, hiked myself up into the big oak tree, and walked slowly out to the edge of the sturdiest branch. My black school shoes were old, the soles wearing thin, and I wobbled back and forth on the knotty wood, my arches doing their best to hold me steady. I watched the man for a minute to make sure he was soundly sleeping—which he seemed to be—and then, holding my breath, I snapped a few pictures. In those early days of taking pictures, I always held my breath until the shot was completed and I was safely hidden from the subject. I found something very mischievous about photography, and this I'm sure is what first drew me to it. It was like getting a piece of something that wasn't mine.

The homeless man's face was framed inside my lens by slender branches spreading across his scarred forehead, dividing his features into countless fragments. My continuous wobbling blurred the photograph slightly, giving each shot a hazy, ghostlike feeling that would become apparent when I developed the film. The following day when this image appeared before me in the Spence darkroom, I fell in love. I was shocked and amazed that I'd managed to capture such a great image. *Me*, who broke everything she touched, and failed at everything she tried. And when the teacher praised me during our class critique I could hardly contain myself.

My photography teacher was a wild, manic woman with long bleach-blond hair and a piercing through the upper cartilage of her ear. She wore black lace-up boots and untucked turtlenecks stained with photo solution. I was in absolute awe. Her photographs were filled with real faces that stared back at me from the pages of her books—bruised faces, sick faces, deformed faces. Faces from the real world. I was shocked and thrilled to find out that another definition of beauty actually existed somewhere other than in my own head.

In the years that followed I became obsessed with photographing people on the fringes of society: the homeless, the sick and dying, addicts, various "freaks." As I grew older the project evolved into something larger and more important, and I began to understand why I was drawn to these people, why they inspired me. Isolated in the frame of my camera, they no longer represented a state of poverty or death or moral degradation, but rather the beauty and strength of their survival. By taking their pictures I was able to preserve this. Capturing these images represented a form of freedom for me, a path to another world I was now learning existed. My subjects were tragic, and I knew that, but there was something beautifully admirable about their tragedy, something much more admirable than I observed in any of the lives I knew.

At some point during my seventh-grade year, after I'd gone as far as I could with photography at Spence, my teacher, with her dry, peeling fingers and awesome mane of bleached, untamed hair, pulled me aside, encouraging me to apply for a summer photography course at the School of Visual Arts.

"You've gotta pursue it, Alison," she told me. "You've got something,"

"I do?" I said, half unsure, half wanting to hear more about this thing that I had.

"Yeah," she said. "You get it. You really get it."

I remember my mother sitting behind her desk the night I asked her for permission to take the course. She looked busy; but after watching her for a few minutes, I realized that she was simply moving papers from one pile to the next, occasionally glancing at one and tossing it into the wooden wastebasket. She'd stop intermittently to sip some wine and stare out her window across the East River.

"That's great news!" she said, lifting her head from the papers. "Too bad it's over the summer."

"It's only for a month, and I can stay alone over the weekends," I pleaded.

"Ali, the swim team will never let you miss a month of practice and meets, and you cannot stay here alone. We spend our summers in Connecticut—you know that." She had stopped looking at me and started pulling drawers out in search of something.

"I can stay at a friend's house if you really don't want me to stay alone, and I hate swim team," I said stubbornly. "I love photography. I'm good at it, Mom. And I want to become better. Maybe go to an art school. My teacher says they have special schools for artistic kids where you can learn so much and have shows and become apprentices to famous artists and I—"

"Alison, it is not convenient for us. There'll be other opportunities—okay? And you're not going to some art school. You're perfectly happy at Spence. End of discussion."

It was then that I lost it. The screams that followed, pealing out of me one after another in ceaseless succession, seemed to be coming from somewhere inside myself I had not yet visited. I could feel their cadence and gravity ripping through me as they rose. I'd tapped into some uncharted territory. Territory I had no place entering. These screams were carnal and animalistic, big, black blubbering things holding the kind of hate and fury that allowed men to kill. And my mother could see it.

She lifted herself from her chair and moved heavily across the carpet until she was standing an inch away from me, staring into my reddening face.

"What's wrong with you?" she yelled.

"You!" I screamed.

Behind me I suddenly heard the clinking of ice cubes against my father's scotch glass. He'd entered the room.

"I can hear you two screaming from the other end of the apartment! What's going on?" he said.

"Nothing anymore. It's all settled," my mother said, walking back to her desk and shifting the zipper of her black wool skirt.

"No, it isn't," I said, more quietly now, not wanting to upset my father, never wanting to upset my father.

"Alison, please, go do your homework," she said.

"I don't care if it is or if it isn't, but I want you to stop yelling. I can't stand it. Talk like civilized people." My father turned from the room.

"Mom, please," I said in a quiet voice.

"Get out of here, Alison. Go to your room, do your homework, and go to bed. I don't know what to do with you anymore. I just don't know."

It was around this time that Ilse was kindly asked to leave. My mother said she had become my "lady's maid," and it was time to let her go. She was given a hefty compensation for her years of dedication to our family and promised monthly lunches and weekend visits. I didn't cry when she left; it took me years to realize how much she had meant to me. Instead, I gave her a forced hug, shouting, "I'll call you," as she went down the elevator shaft and knowing very well that I wouldn't.

My mother sent her Scottish smoked salmon and hams during the holidays and called her on birthdays, insisting that this was the polite thing to do. I didn't see her again until I was twenty-seven. Her knees had given out and she'd had an operation. I went to visit her in the hospital. At first, I didn't recognize her. Her limbs and torso seemed to have shortened, as if she'd shrunk into herself, and her thick black mop of hair was now graying and thin. But as she spoke from the hospital bed, her leg strapped into some movement machine meant to keep her knee in constant motion, her blue-and-white quilted hospital gown scrunching as the machine moved, the tender and amiable Ilse I remembered emerged. In her soft voice she told me about the children she'd cared for over the years, detailing the childhood of each boy and girl with such care and precision that Ilse's distinct sensibility came shinning through. And as I got up to kiss her goodbye, trying to hide how much it hurt to see her like that, she grabbed my arm and said, "We had many good times, didn't we?"

I nodded, smiling through the lurch of my heart, and kissed her again before saying goodbye.

After Ilse was let go, my mother started drinking heavily, polishing off three bottles of white wine a day. I kept track, diligently watching the third drawer of the refrigerator that was always lined with deep green bottles of Pouilly Fuisse. Each evening her cocktails appeared earlier until the glass was a permanent fixture in her hand. My mother's drinking, though always present, escalated when my wild ways evolved from childish spunk into volatile teenage anger. Her behavior began to frighten me. My father and I had a code for this behavior. *Mom's gone to the crazies*, he'd say, or *Did the crazies get her again last night?* Everything involving emotion was encoded with him. He never said *I love you*. When I said it to him, he'd only answer back with *ditto,* as though even the utterance of such words made him too vulnerable.

After that initial explosion about the photography class, my fights with my mother became more frequent. I no longer kept things to myself. I no longer pretended I wasn't mad. I felt like I'd discovered a weapon I didn't know I'd had available to me, and I could see the effect it had on my mother. I liked it. I could crush her under my words. I could send her spinning into a state of total bewilderment until she couldn't utter a coherent sentence and was reduced to nothing but a blubbering infant. This made me feel powerful.

Our fights usually happened late in the evening after my father was asleep. He'd requested that extra humidifiers and a white-noise machine be brought into the room. The fights were gross and loud, full of ugliness and loathing and despair. Sometimes they'd turn physical. We'd fall to the ground yanking at each other's hair, scratching or squeezing each other's arms. I think there came a point when words were no longer enough.

In early June, a few weeks after seventh grade ended and school let out, we left for Connecticut. The trees were green on Sutton Place; children wore shorts and sundresses and carried juice boxes as they held their nannies' hands or rode shiny, honking bikes on their way to the park. My parents watched as the doormen packed up our station wagon, its air no longer light and flowing but rather divided into three separate units, unshared and motionless. By early July, I had been kicked off the swim team for arriving late to practice, and spent my days in the Connecticut woods getting high on marijuana with older friends from the country club, none of whom were on the swim team.

I smoked my first bit of pot that July out of a long, purple phallicshaped bong. During the day we'd lie around the club pool, eating cheese fries and playing Marco Polo and begrudgingly dragging our sun-drenched bodies from tennis clinic to golf clinic.

My father still insisted that I practice my tennis game with him at least once a week, and because that fierce desire to please him was still alive inside me, I did. But the problem was that I was no longer impressively talented for my age. I was out of shape, and I didn't have the muscle strength to lob the ball like I once had, and I'd lost my slice and serve and spin shot.

"You have to run for the ball, Alison!" he'd yell across the court.

"I am, Dad," I'd scream back.

"You're losing zero to five. You used to score better than that when you were seven."

"Can't we just play for fun?" I'd plead.

"We are playing for fun. It's fun to win—remember?"

"Not if you're screaming at me the entire time."

"Well, if you play better, I'll keep quiet."

That was all he ever said that summer when we played—*harder, faster, where's your slice, keep your eye on the ball, hustle, hustle*. When we left the court, I'd be near tears, and he was almost always infuriated, fuming with frustration as he shoved his racket into its case.

"She'll never amount to a decent player," he said to my mother one afternoon as we entered the house.

"Why not?" she said, looking up from a magazine, wiping a wisp of hair from her cheek.

"She's no good, and she won't put in the time it takes to get better. You weren't any good either. None of you women are any good at anything that matters. I'm canceling the clinic."

"Fine with me," I said. He looked at me, disgusted.

"You're a waste, a total waste," my father said.

"Will, stop it."

"You are, too. Both of you are. I'm sick of you both." He banged out of the room.

My mother and I looked at each other for a minute. Then, shrugging, she wandered toward the bar with an empty plastic cup in her hand. I charged up the stairs two by two, locked myself behind the bathroom door, and feigned a shower while smoking a meticulously crafted joint.

I didn't acquire a fondness for the high marijuana gave me until much later in my teen years. At that point, my desire to smoke pot came from a misguided understanding I'd formulated along the way—belief that drugs were the gateway to the artistic life I so deeply yearned to have. When I returned to New York that fall, my friends and I began making weekly trips to a small store hidden behind scuffed bulletproof glass, a bodega in Harlem. Hanging from the top of the door was a small rusted bell that jingled every time someone walked in, and to either side were shelves holding random household products, canned foods, and sodas that all appeared to be years past their expiration dates. I'd learned about the place from some older boys at the Eighty-sixth Street pool hall, where we'd recently begun spending our Saturday nights. Once inside the bodega, we'd head to the back right corner where a window barely big enough for a kitten to crawl through had been cut out from the wall

above what was once a legitimate countertop. You could still see the outline where the register once sat. It was here we would knock three times,
place a twenty-dollar bill through the window, and watch as a perilously
thin hand with branching blue veins crept, like a dying rodent, toward
us. The hand would take the bill, retreat, and return bunched around a
small ziplock bag of pot.

Back on the street, a rush of air from the subway grates would blow
our plaid uniform skirts up, and the men who hung around outside the
shop would scream, *Hey sexy*, making lewd gestures with their fingers and
tongues. Oblivious to any danger, we'd run haphazardly into the street
screaming and giggling and lifting each other's skirts into the air, until we
saw a free Gypsy cab and jumped inside.

This wasn't unusual behavior for me or my girlfriends. My early
teenage years were filled with many reckless nights of guiltless taunting
and teasing, random mayhem, and total disregard for limits of any sort.
We roamed the streets of New York in a constant state of degeneracy, and
it's a wonder to me that our reputations as refined, educated young ladies
held on for as long as they did.

6.

ICE CREAM

I was caught in possession of marijuana during the first semester of my eighth-grade year. I woke up one morning and noticed my closet door ajar and the hanger for my winter parka empty. My mother had taken to snooping around my room while I slept. She always left clues; drunks aren't exactly meticulous about covering their trails. I often found a wine glass or a crumpled tissue on my shoe shelf, her reading glasses on my bureau.

That morning, she was sitting on a stool at the end of a long wooden island in our kitchen, a cigarette hanging from her mouth, a Bloody Mary on the counter beside her. The help was off. It was a Saturday. She looked sepulchral in her thin white bathrobe, her dyed blond hair falling in a kind of dead tulip wilt around her sallow face. When she saw me standing there, she yanked the cigarette from her mouth and dropped it into a cup of water that already held half a dozen butts. She was still pretending, still trying to pull off a facade, even to me.

"I know you smoke, Mom," I said.

"I don't smoke. I quit years ago. I'm just upset this morning."

"Can you give me the pot back? It wasn't mine. I was holding it for someone," I lied lamely.

"I don't believe you, Alison. And you are not getting it back." She stood up and walked out of the kitchen.

I followed her into the living room where her flower arrangements were spilling from various vases on the corners of tables. She was adamant about having fresh flowers in every room of the house. We began yelling at each other.

"Shut up! Just shut up!" she finally yelled over me. "We can discuss this further in family therapy. I've made an appointment. We're starting next week."

"Fuck you," I snarled. "I'm getting a lock."

We attempted family therapy the following week. I met my parents at 3:30 PM on East Sixty-ninth Street in the waiting room of Dr. Azara's office. My mother was in the bathroom, "tinkling," as she called it, and my father was reading the paper.

"Hi, Dad," I said, sitting next to him on the couch. He and I were still friends from time to time.

"What are we going to do about this nonsense?" he said. "Your mother dragging us here. Are you using drugs?"

"No, Dad, I was holding them for someone. I told Mom, but she doesn't believe me. I don't want to be here either."

"I didn't think you were stupid enough to fool with that kind of thing."

My mother came out of the bathroom. Freshly glossed lips broke into the smile I knew so well. I turned my eyes to the carpeted floor. We sat in silence until Dr. Azara came out and greeted us.

"Nice to see you again, Rosemary." She shook my mother's hand and turned to my father and me. "This must be the rest of the family."

Then she motioned us into her office. Dr. Azara had one of the straightest gaits I had ever seen, and her pointy features seemed to have been carved with an X-Acto knife, her pageboy haircut mirroring the same razor lines. She had an attractive face but the hair and the features were too severe for someone meant to evoke comfort and safety. We followed her into the office and sat down. The room was dark with maroon walls and leather furniture; the radiator ticked behind me. Mother sighed, pulling a tissue from her sleeve. Dr Azara cleared her throat and came straight to it. "Are you using drugs, Alison?"

"No," I said. "I told my mother I wasn't, but she didn't believe me."

"Then how do you explain this?" my mother said. She drew the ziplock bag of pot from her Hermes purse as if she was an actress in a second-rate detective movie.

"I told you. I was holding it for someone who didn't have pockets. Not all coats have pockets, you know?"

Dr. Azara cocked her head and squeezed her eyes into the center of her face as if to acknowledge the ridiculousness of my excuse. "Okay, let's move away from the drugs for a bit. I understand you're not getting along with your mother and father lately?"

"We get along fine," I said.

Dr. Azara turned to my father. "Mr. Weaver, how do you feel about your relationship with your daughter?"

"It's fine. My wife dragged me here. I don't believe in therapy. Any problems that we might have we can work out on our own." But I could hear his voice cracking.

"Alison never speaks to us, Will," my mother said. "She's breaking school rules left and right, she doesn't obey her curfew, and she's hanging out with people who smoke this!" She lifted the bag of pot again.

My father stood up. "This lady doesn't need to know our business, Rosemary. Alison had all A's and B's on her last report card, and she talks to me plenty. There is no reason for me to be here." He walked out of the office.

"I'm going too," I said, jumping up after him.

We stood for a minute under the awning. It was midwinter, and I could see our breath leave our mouths and trail past each other. My father wore a long olive-colored wool coat with a maroon and green muffler and a matching English cap. In his hand was a hard leather briefcase monogrammed in gold that he swung nervously.

"Should we go get ice cream?" he finally said.

"Sure," I said, and as we started up the block to Serendipity, he kicked me in the behind the way he used to when I was little.

We shared a frozen hot chocolate, Serendipity's signature dessert. He talked about problems with the cattle business out West, dishonest workers who were stealing from us, a ranch manager who was fixing the books. I listened attentively, as I always did when he spoke. Perhaps I knew a time would come when one-word answers were all anyone could get out of him. After we'd finished all we could eat of the gigantic bowl, he quieted and stared down at the deflated whipped cream and melted, soupy chocolate.

"That's one thing that will never let us down," my father said.

"It always holds up, doesn't it?" I said.

"Yes, it does."

Then, as we were putting our coats on to leave, he said, "Do you think the crazies have gotten Mom badly?"

"Pretty badly," I said. I paused for a moment, then continued, "I don't get why you don't just divorce her."

He lifted his head and looked me in the eye. "I'm too old for a divorce, Al," he said. "Listen, don't let them get you, okay?"

"I won't, Dad," I told him.

This was my father's way of reaching out. He was trying to do his fatherly duty, but I could see the apprehension in his face. He didn't want to accuse because he didn't want to believe it himself, but he must have noticed that I'd been behaving differently, badly, growing farther and farther away from the little girl he loved so much.

My father and I would have other moments like this in the years to come, moments of reconnection shadowed by the understanding that what we once shared was unrecoverable. My father liked life to run according to plan, and when people strayed off course, he wasn't good at helping them find their way back. Often the minds of the most intelligent people are intricate, thorny places, but my father's exceptional mind was "brain" in its simplest form: everything was as it seemed. He had no time for regrets, doubts, or insecurities, and couldn't comprehend why others did. He smoked a pipe for twenty-eight years, then one day decided to quit and did, just like that, no questions, no patches, no support group. To him psychoanalysis was complete nonsense.

My mother never forced us to return to Dr. Azara after the day we

Mom

left her in the office. She must have known it wouldn't work. Now, years later, I give her full credit for trying to heal—or at least to recognize and investigate—what had gone wrong with our once relatively happy home. She only failed us in refusing to acknowledge her own responsibility, her own contribution to it all.

7

ONLY RED

There was one psychiatrist who I was fond of growing up. His name was Dr. Milton and his office was on York Avenue, located directly across the street from New York Hospital. I had been sent to him during my seventh-grade year at Spence after stealing a tape of Beethoven sonatas from the music room.

His office was professional, black and brown with a large wooden desk, a Sharper Image lamp, and pictures of his two girls lining the office windowsills. Two blond sisters, arm in arm in front of an elaborately trimmed hedge, on horses, on skis, in bathing suits cuddled next to their father in what I supposed was their beach house. Dr. Milton wasn't generically attractive. His hair was graying and curly, and he had a potbelly and fleshy jowls hanging below his chin, but I was oddly drawn to him. He was wickedly smart and sassy, and gave off a slight air of pompous confidence that sucked me in.

I loved the way he leaned back into his cushy leather chair as he thought over things I said. He'd cross his legs, tap his slim foot encased in its tasseled leather shoe and lift his bushy eyebrows. The movement of his eyebrows seemed to create a camaraderie between us that was most welcomed at that point in my life. But most appealing was the way he smoked, as if each cigarette was of the finest tobacco he had ever tasted. He smoked with passion, inhaling deep, long drags and blowing the smoke slowly into the air as if parting with it was profoundly troubling.

It seemed to be the way cigarettes should be smoked, and I began to take the process of smoking equally as seriously.

"All right," said Dr. Milton, pulling out an ashtray from his desk drawer at the beginning of our first meeting.

"I can smoke in here?" I asked.

"Sure," he said.

Dr. Milton never made me discuss things I didn't want to. We talked mostly about my friends. He let me gossip about them, vent my frustrations. Sometimes, I'd tell him about fights I'd had with my mother but as the years progressed and they grew uglier, I began to reveal to him only the watered-down versions. I never divulged the true nastiness of them, never spoke about her wine glasses that smashed against my bedroom wall, her nails that left red furrows along my arms. Usually our sessions consisted of friendly chit-chat. Sometimes he even talked to me about his past of famous clients like Janis Joplin and Stevie Nicks.

"Do you drink, Alison?" he asked me one day, after I'd been seeing him for nearly two years.

"I have," I said.

I knew what he meant, but I was reluctant to answer the question, suspicious that my mother was behind it. I had taken my first sip of alcohol at five years old from the metallic-tasting lip of my father's beer mug, a tin one with a glass bottom. I loved to lift it up to the light and look at the ceiling through the yellow foamy liquid. The taste of beer revolted me, but still I often begged my father for a sip because the look of mischievous pride that came across his face when my mother's back turned was well worth the momentary disgust.

The first time I got drunk was on a cruise ship in the middle of Alaska. I was nine. It was New Year's Eve, and champagne was flowing, and I was running around the casino with some older girls. I played the slot machines and won thirty dollars, and then we all went into a cabin and popped open the complimentary champagne bottle sitting on the mini bar. It didn't take more than two glasses. I awoke in the morning with my party dress still on and my bedsheets untucked.

My head throbbed, and my eyes ached and my mouth was dry and sour.

But it was the summer of my fourteenth year when I really began drinking. I can't recall the first time because in the early years nights all consisted of the same thing: sitting around in parks or behind the locked doors of a friend's room, swigging wine coolers and chain smoking. Teenagers don't drink because they love the taste of alcohol; they drink to get drunk, and that's what we did. I was by no means an anomaly among my friends. We drank; we got drunk; we giggled and laughed and made a few prank calls; and we stumbled home.

"But do you drink on a regular basis?" Dr. Milton asked.

I drank most every Friday and Saturday night unless I was stuck in Connecticut with my parents. I knew my mother had smelled the alcohol on my breath and noted the sloppiness of my movements. I was sure she was behind these questions.

"Anything you tell me is confidential, you know," Dr. Milton said, sensing my apprehension.

"Did my mother ask you about my drinking?"

"She's concerned."

"But you can't tell her anything I tell you?"

"Unless it puts your life or the lives of others in immediate danger."

"Doesn't drinking do that?" I didn't want to leave him any loopholes.

"If you tell me you're planning to drink a bottle of vodka and take your car for a joyride, I will tell your parents. Otherwise, no, it doesn't," he said, fed up with me.

"Yes, I drink on the weekends," I said.

"A lot?" he said.

"Some," I said, knowing that I drank a lot more than some.

"Do you spend time with your parents to try to help them understand you?" he asked, changing the subject.

"We used to have a designated family time every night from 7:30 to 8, but it rarely happens anymore. I just don't have anything to say to them."

"Well, couldn't you think of something you wanted to tell them?"

Dr. Milton grabbed the pregnant ashtray, dumping it in the garbage for me.

"I could, but the thing is I really don't care anymore. It's not supposed to be work," I lowered myself in the comfy leather chair.

"What do you mean?"

"Talking to my parents, living with them, it's not supposed to be work, so why should I work at it? I have good friends. I don't need my parents to be my friends too—not the way they are." Slightly uncomfortable, I glanced at my watch.

"Time's almost up, but, well, it seems to me that you do need them. Or else why would you be complaining to me?" He leaned in, awaiting my answer.

"They don't understand me, Dr. Milton. I mean, *I* don't even understand me, but they aren't even willing to try. They have a block. It's like they want me to be the color red. Only red. I want to be blue, or maybe even yellow or a mixture of both, but they won't even consider that. It's red or nothing in their minds, and if I'm blue or yellow, then something is wrong. I need help, and I'm shipped off to another psychiatrist. That's how the whole world is." I grabbed a loose Marlboro from my shirt pocket.

"They're just trying to help you. They know you're unhappy."

"I am not unhappy. I'm just frustrated," I said.

"Well, you need to communicate that frustration, Alison. You can't just be angry and slam doors on them. You have to tell them what they are doing wrong," Dr Milton insisted.

I lifted a Zippo lighter from his desk and lit my cigarette.

"And you smoke way too much," he said, leaning back in his seat. "It's not healthy, kiddo."

8

THUGS IN BROOKS BROTHERS

Eighth grade brought my first Goddard Gaiety dance. This was a traditional event that students of the Upper East Side private schools looked forward to as soon as they could talk. It signified passage into the grown-up society much in the same way a coming out party did. We had seen our parents play socialites since we were born, seen them point out their friends or themselves in the fashion columns of *W* magazine, *Avenue*, or *Quest*. Now it was our turn to schmooze and impress and brag. Time to claim our place in high society New York. The boys would get special suits tailored and wear them with panache and pride, their Harry Winston cuff links and French-cuffed shirts peeking out below the arms of their jackets, believing this is what made them men. The girls would have spent weeks picking out the perfect dress, velvet or wool, never taffeta or polyester or spandex. They'd spend hours at the hairdressers, maybe get a makeover at Bloomingdale's and practice walking in their new satin high heels for weeks before the event.

It was January, and the city was still lit up with Christmas decorations. Light snow had begun to powder the sidewalks, and the branches hanging off trees that lined the curbs were weighted with layers of white dust forcing them to slouch toward the ground. Doormen stood at attention under green and navy awnings and people swam in and out of one another with shopping bags and baby strollers. My friends and I had rented a limo. We drove up and down Park Avenue, sipping champagne and sucking on sloppily rolled joints and otherwise pretending we owned the world. The boys from St. Bernard would be at the dance: Addie, Mike, John. I was jittery, nibbling my manicured nails in between cigarettes as we downed glass after glass of bubbly. We pulled up to the ornate Catholic Cathedral on Park Avenue where Jackie Kennedy's memorial service had been. Outside the tinted windows, boys in suits, dressed like Wall Street businessmen, were filing up the stairs, the popular ones mingling with the girls, the shy ones huddled in groups trying their best to remain invisible.

I am aware that there was a contradiction present in my behavior here, and I acknowledge that often in these years there was. It is human nature that given time people will usually succumb to their environment, however adamantly opposed to it they once were. In the back of my mind, I still hated these frivolities and games; I still knew that it all meant nothing—the emptiness still terrified me. But this was the only world I knew, and I hadn't yet figured out how to break away from it. I wanted something to change, but I didn't know what, so I sat in the limo, and drank the champagne, and wobbled up the cathedral stairs in my new two-hundred-dollar heels with everyone else.

With friends from my teenage years

The dance room was decorated extravagantly. The walls were dotted with crushed tissue paper: pink, purple, green, red. Balloons floated above us, the ends of their strings gently oscillating as boys jumped to graze them with a fingertip. The church basement was lined with tables of food and drink, each with two punch bowls on either side, thin orange slices floating on the red surfaces like lily pads. A deejay stood on the stage, alternately sifting through crates of records or fiddling with knobs on the turntable in front of him. Most of the students had just returned from winter break and were endowed with tropical tans. I had spent the two-week break in the Bahamas at the Lyford Cay Club, and I wore a skimpy spaghetti-strap dress to flaunt my dark color. We mingled by the punch bowls chatting about our wild vacations, complaining about being back at school and quoting the newest rap songs.

By the end of the dance, I was seeing double. I tried speaking to Buckley, a St. David's boy whom I had developed a crush on a few months ago after we'd French-kissed during a game of spin the bottle. Shirt untucked, jacket off, probably slung over a chair somewhere in the corner of the room, he stood arrogantly, tossing his sweaty blond hair from his face. He was slight, around five foot five and no more than one hundred pounds. On his pinkie was a thick gold ring, a signet ring with a family crest tastefully stamped on the front. His friends were cruising the room in search of ripe and ready girls, developed breasts to squeeze, or fistfights to prove their machismo. Darting among the adult chaperones, they rated the girls from one to ten slapping each other on the back and doubling over in laughter when a girl received a negative rating. These boys may have been the young elite of New York City—well-groomed, well-spoken, well-educated—but in truth, they were no better than thugs in Brooks Brothers suits.

I wanted to think Buckley was above all this, though he was probably aware of my obliterated state and figured the odds of hooking up with me were in his favor.

"You've got a nice tan," he said, as we stood nervously by the punch bowl.

"I was in the Bahamas," I said.

"I was skiing in Aspen. I won a silver medal in the Nastar downhill."

"What was your time?" I asked, relatively confident that mine was better.

"I forget," he said. "My brother is on the varsity ski team at Choate. I plan to go to boarding school too but probably not Choate because my grades aren't good enough. Maybe Millbrook or Berkshire. They've got a good ski team, and I know the captain because he's my friend's older brother. Are you going to boarding school?"

"I haven't thought much about it," I said, growing bored with the conversation.

"Do you want to dance?" he asked.

"Sure," I nodded as he grabbed my hand, and we wandered into the pit of sweaty, jumping bodies in the center of the room.

As the months progressed, these nights began to grow more wild. We stopped going to the school dances. Instead, we hung out at people's houses when their parents were out or away for the week. We drank heavily, chain smoked, touched each other.

Near the end of eighth grade, we threw a private party. An unpopular girl named Alice Nesbitt was moving from her West Side walk-up to a duplex on Park Avenue; her father had recently been promoted. Alice seemed to have been imported from the 1950s; a cover girl for one of the elite finishing schools our mothers used to go to. She had a high blond ponytail made up of perfectly curled hair like the ribbons on a fancy birthday present. Her nails were always manicured with clear polish, and she wore fuzzy pastel sweaters and Mary Janes with white lace socks.

I asked her if we could use her old apartment, knowing her desperate longing for acceptance would force her to say yes. She agreed. The apartment was empty. No furniture anywhere, just some broken boxes we used as seats, empty tape rolls, and old Styrofoam. I played bartender, whipping up random mixtures of vodka, rum, Kahlua, ginger ale, orange juice—whatever people had brought. By ten o'clock the place was so crowded you could barely move.

At some point during the evening, I decided that I was ready to lose my virginity to Buckley. I remember watching him from the kitchen as he moved smoothly through the crowd, high-fiving his buddies and pecking girls on their cheeks.

"Hey, Alison, the party's happening," he said upon reaching me.

His spearmint breath heated my cheek, and I felt his pinky slide resolutely along the inside of my bare thigh. A jolt of something I'd never felt before shot into every pore of my body, and I leaned back, grabbing a bottle of vodka from the edge of the counter. I poured us each a shot, and within the hour we were behind the locked doors of the hall bathroom, peeling each other bare. I unbuttoned his shirt, and the rugged odor of boy infused the tiny bathroom. He touched my face, traced the outline of my jaw. I rubbed the thigh of his khaki pants. His hands dropped to my neck, then cupped my small breasts. I had spent many previous evenings in various bedrooms across the city, slithering around under sheets with various stoned boys as Pink Floyd played on the stereo.

But that evening, I was paralyzed with the knowledge of where this was going. Our interaction turned mechanical and the tingling that had encompassed me earlier went dead. I couldn't get out of my head and back into my body. All I could feel was Buckley's long, slimy tongue licking the roof of my mouth, the sides of my cheek and throat, his sweaty hands squeezing my breasts—not gently as I had imagined he would, but as if he were milking them. Suddenly the act became too real. I kissed harder, trying to focus on the sensations, but all I could feel was my sweaty back thrust against the cold mirror over the sink and the faucet jabbing into my waist.

We came very close to having sex that night, but eventually we were interrupted by rowdy knocking on the other side of the door. Buckley's pants were bunched down around his ankles, and the soft head of his underdeveloped penis shyly popped through his pink Polo boxers. My skirt had been pushed up around my waist, my top was off, and his fingers were inside me.

"Be right out," he called, and then turned to me and shrugged his shoulders.

"Oh, well," I said, secretly relieved.

The party died down around 1 AM. The girls and I left the apartment stoned and starving. Someone suggested we stop at a deli, so we headed toward Broadway where the city still pulsed and flashed. The stiff leather straps of my heels had formed a half anklet of blisters above the buckle, and I could feel the wetness of blood on my skin. I must have been swerving slightly because I noticed a couple on the other side of the street look over with disapproving stares. They'd been laughing, but when they saw me they stopped, turned their heads, and walked briskly on, no longer in laughter, but sort of nuzzling into each other's shoulders, hoping to escape whatever unpleasant thoughts a fourteen-year-old girl stumbling down the street in nothing but a crushed velvet black mini skirt and a tight halter top might evoke. It was November, but I wasn't cold.

Neon letters flashed brightly above, beautiful faces smiling down on me from billboards like one-dimensional angels promising salvation. Naked glittering bodies with lips, full and cherry-red, gently blew kisses to lonely men, luring them into triple-X booths. Some kids were break-dancing in the middle of a cheering crowd of tourists, clapping and bouncing around without rhythm or thought, while beside them homeless men sang songs or told jokes for money. They didn't seem sad or hungry but alive and full of joy. Nothing around me seemed sad. It was all too colorful and exuberant to possibly be anything but happy. I could hear laughter rising and receding in the distance.

I had fallen behind the girls. Making out the movement of hair and sequined tops and bare arms and legs, I staggered toward them. Then they were inside the deli. I could see them through the glass window: five unruly teens, drunk, high, and screaming under the fluorescent beams of deli light, capering down aisles and tossing neon-colored candy back and forth. Starbursts, Skittles, Caramellos, Blow-pops, Push-pops, Ring-pops, Sour Patch Kids, 100 Grands, Milky Ways, all glowed incandescently like the flash of summer fireflies. I entered. Someone threw the cashier a fifty-dollar bill and told the man to keep the change. The rainbow-colored walls were blurred and spinning slowly but sickeningly, like the perpetual motion of a carousel, and the girls were speaking to me. I could see their lipsticked mouths moving and their shiny teeth gleaming, but it was like watching a teen movie on a muted television set.

"Alison," I finally made out. "Snap out of it. The party is over. We didn't know where you went. Let's go."

The deli walls were spinning faster now, my stomach heaving up and down like it used to when my father pushed me too high on our old hammock in Connecticut. I was unable to get myself together. I had been drinking at a strong and steady pace for the last six hours. I couldn't speak or see clearly, and when I noticed the Plexiglas tubes filled with hundreds of brown pellets lined neatly against the deli walls, I lurched toward them across the shiny floor, somehow believing that by exerting control over these coffee beans I would regain control of myself. My lanky, bangled arm reached out, and I watched my hand yank up the sliding doors, one after the other, jamming them open with various candy bars. And I stood watching contentedly as thousands of tiny coffee pellets cascaded from the mouths of the tubes pouring across the floor until my feet were nearly submerged.

The girls were now yelling, pointing, yanking each other this way and that.

I heard nothing. Just hard, tiny pellets rattling onto the linoleum floor like hundreds of Lee press-on nailed hands tapping at windows across the city. My eyes locked on the falling beans, and I remember thinking they looked like a waterfall that might feed the rivers of Hell: thick, dark, and relentless. Someone jerked on my hand, and I felt my shoulder joint pop as I was dragged outside onto the sidewalk where the other girls were screeching, flagging down taxis. I was on the ground, rolling around on the sidewalk, falling alternately into fits of delirious laughter and tears. Looking through the deli window, I saw the owner, shiny with sweat, sweeping the coffee beans into a pile at the center of the store. For a brief second, I wondered when I had devolved into such an awful mess of a human being. In an unusual display of compassion, two of my friends carried me around the corner, propping me against an old townhouse gate and trying to force pretzels down my throat.

The next thing I knew I was alone in a cab, and the driver seemed to know where we were going, even though I had no recollection of telling him. The stiff, cracking cab-seat leather dug into my bare thighs, and specks of dried vomit covered my skirt and legs. I noticed a cigarette in my hand that I must have taken out to light, but apparently I hadn't managed to follow through with the plan. The night outside had turned dark and silent. Central Park whizzed by on my left; dim street lamps highlighted eerie shadows in the darkness under benches or in corners of piled leaves and trash. The playgrounds with their metal see-saws creaking in the wind-like old wind-up toys, and the empty, twisting roads leading under bridges into black, all seemed to belong to a place where every imaginable crime was committed. The formal, stone-cold architecture of the Natural History Museum loomed ominously on my right. My toes curled tightly into themselves, the toenails scraping the inner soles of my shoes. I ran my tongue across my chapped and bitten lower lip.

I lay down on the cracked leather and without thought I began to bang my head into the taxi's metal door handle like some sort of autistic child in the midst of a fit.

"Hey, lady, what you doing back there? Stop it," the cab driver yelled into the air, turning up the volume of the radio and accelerating.

"Sorry," I whimpered, placing my hands over my now-throbbing head. I opened my eyes wide and looked up through the smudged window for those one-dimensional angels but they were gone. It was just the dark city sky with a few indecipherable yellow shapes.

What seemed like hours later, the taxi stopped in front of the green awning of my apartment building. I felt in my pocket book among the half-smoked cigarette butts and tic-tacs and old receipts and pistachio nutshells until I unearthed some crumpled money. The lobby was empty except for the night doorman dozing off on the bench. I rode the elevator to the sixth floor, and found my mother pacing the foyer in her lace nightgown, white wine in hand. We stared at each other for a minute, then I ran to the bathroom, slamming the door behind me and collapsing onto the tiled floor.

She knocked forcefully from the other side—*thunk, thunk, thunk*—her hard, demanding knuckles thudding against the door's glossy cream finish. I didn't answer. The door swung open, and she came at me, her face filled with rancor and her voice unraised but scathing. She grabbed my shoulders and began shaking me.

"What is *wrong* with you? Where were you? God, I thought you were dead!"

"I told you I'd be late," I cried.

"Why are you doing this to us? You're going to kill your father. He's too old for this!"

"You married him! It's not my fault!" I cried, pulling away from her.

In slow, unsteady movements her half-masted eyes traveled across my body, noticing it was soaked in vomit. Leaning against the glass shower door, she slid to the floor, sitting with one leg sprawled to the left, the other bent, and her elbow resting on her knee. Her face had softened. Now her head hung low, and tears fell from her eyes, splattering in an uneven circle on her bare thigh. It was peculiar to see her in that position. My mother didn't sit on floors. Even when I was young, and she visited me in the attic playroom, she'd pull up a chair and lean over her knees to reach me. Seeing her there on the floor, she seemed so vulnerable that for an instant my heart dropped and I felt sad for her, for us.

But only for an instant, and then I was vomiting into the toilet for the good part of an hour as my mother sat beside me on a plastic stool. She wet a washcloth, gathered my hair into a ponytail and dabbed my chin, maternal and concerned, as if she'd come to terms with the circumstances. I apologized profusely, promising things would change, begging her to forgive me as people do when they're sufficiently obliterated.

"I love you too, Ali," she said over and over again as she tucked loose strands of hair behind my ear. "Everything is going to be just fine, like the old days," she said, and I nodded, crying, wanting to believe her.

9.

NOT EVEN A LITTLE BIT SAD

A week after the coffee bean incident, I was expelled from Spence. One of the girls I was with that night had been wearing a Spence jacket, and the deli owner had called the school and demanded that something be done. I was blamed not only for the coffee beans but for providing the alcohol as well.

Part of me knew that this outcome was inevitable, that the relationship between Spence and I had always been predisposed to disaster, and in order for it to succeed, it would have required meticulous care and attention that I was unwilling to give. Realistically, I could never have been the student Spence required of me—nor did I want to be. But on the day they expelled me, I still begged and pleaded with them to take me back. I swore to that headmistress I would change, swore I would be well behaved and get straight A's, stop drinking and smoking pot. I would focus, I really would. But she sat unbendingly behind her cherry-stained wooden desk, strings of pearls sliding around her neck as she nodded her head coolly to my pleas of forgiveness, her austere, dictatorial face unmoved by my tears. Within minutes she had escorted me to my locker to collect my belongings. Through the classroom door in the hallway, I saw Mrs. Wilson's Ancient History class being taught, and my friends diligently at work on their immaculate notes.

I walked home through Central Park. The sky was gray, but the

rain had ceased, and parts of the puddles were solidifying into thin layers of ice that I shattered with my heels as I walked. I passed the children's zoo I had once frequented with Ilse, and I heard the carousel in
the distance. The air smelled of hot dogs and relish. Children ran about
on the huge rock across from the zoo, their nannies in white uniforms
chatting with each other on the park benches. They complained about
this mother and that father, and occasionally yelled an order out to
their charges. Further down the path, I found a small trickling waterfall
with pennies at the bottom. My gaze became lost in the ripple of water,
the tree shadows skipping across half-frozen chunks of ice, and I wondered if any of the wishes those pennies symbolized had come true. It
began to drizzle again.

Throughout the years vario diagnoses had washed over me like words
reflected from the light of a projector. There was *conduct disorder* and
anxiety disorder and *dysthymic disorder*, the words battering at me incessantly until in my teen years I'd started lying to the psychiatrists, pretending I was happy. Dr. Milton was the only one I was even moderately
truthful with, and he was also the one who suggested I had manic depressive disorder. That my prevailing wild behavior was characteristic of
a manic episode: excess energy, fast-talking, grandiose ideas, overconfi-
dence about certain things, racing thoughts, risk-taking behavior, easy
distractibility. For most people these were the atypical episodes, but for
me they were the prominent ones. He said I protected myself from the
depressive lows with substances, but sometimes the breakdowns were un-
avoidable, and I'd crash and become hysterical and hurt myself.

"Do you hurt yourself, Alison?" he asked me once during a session.

"What do you mean?" I said.

"Well, you know—purposely hurt yourself. Cut yourself or punch
your hand into a wall or bang your head."

"Not really," I said, uncomfortable with the question.

"So you *have* hurt yourself," he pressed.

"Well, hasn't everybody?"

"No, I don't think so," he said. "Listen, Alison, have you ever heard
of manic depression?"

"Sure," I said.

"I think you might have it. An uncommon form of it, but it nonetheless."

I remember squirming in my chair as this sentence left his mouth. The word *depression* was like a tiny electric shock to each of my cells. Every inch of my body had an aversion to this word, and I laughed out loud at Dr. Milton.

"It's not funny, Alison. If you do have this disorder, we can get you the right pills and make life a hell of a lot easier for you," he said.

"No thanks. I don't have that disorder, and I'm not taking pills."

I'd made up my mind. I wasn't manic-depressive. I wasn't depressed. I wasn't an alcoholic. I wasn't even a little bit sad. None of it, I wanted none of it. And I made what I thought to be a rational decision to block it all out, shut it up, numb it, dilute it, deny it, mask it; call it what you will.

I'd like to say I had no control over my behavior in those years, that I was swept away into the tumultuous storm of depression and addiction. It would be nice to place the blame upon those ambiguous conditions, but I don't believe that was the case. I leapt out of reality deliberately, eyes open and with total awareness. It didn't require much effort, just a flick of the mental switch. Imagine it as hooking your soul up to a permanent IV line of Percocet.

Of course, I didn't understand any of this then—that's how it was in those years. I sometimes did things, and a moment later I would wonder why I'd done them. Perhaps, if I'd listened to Dr. Milton, taken medicine to even me out and teach me to think and slow down my racing thoughts and rapid speech, there would have been a different version of my life to tell, but at that time I adamantly believed that he was wrong. I believed I knew depression, and this was not it.

The afternoon I was expelled from Spence, I heard my mother come home around five. Normally she came straight to my room to check if I was diligently at work on my homework, but that day she didn't even acknowledge that I was home. I knew then that the school must have contacted her. Instead, she went straight into her bedroom and clicked on the television. Soon I could smell the smoke from her cigarette and hear her lonely, monosyllabic coughs through the buzz of the newscasters. An hour later my father arrived, and I heard his heavy briefcase drop onto the foyer floor. He walked to the coat closet in the hallway and removed a hanger from the rack. I visualized him taking off his gloves and cap, unwrapping his scarf and hanging his overcoat on the hanger. I heard my mother shuffle down the carpeted hallway and whisper something to him. Usually after work he would have gone right to his dressing room and soaked in a hot tub for close to an hour, but that night he followed my mother to the living room, which was located on the other side of my bathroom. I put my ear to my wall.

"What are we going to do now?" my father said. "What is going on with her?"

"I don't know. Where is she going to get in at this point? No place is going to accept her."

I heard the bar door screech open.

"I just don't understand it," my father said somberly.

"We certainly can't send her to public school in New York, which is what she's going to suggest," my mother said. "That is absolutely out. She doesn't need to be free and loose in New York City thumbing her nose at us the whole time."

"There's always boarding school," my father said.

I decided to put on my bathrobe and go in. Because they were totally immersed in their conversation, I was able to stand in the doorway unnoticed for few minutes.

"No decent boarding school is going to accept someone with expulsion on their record, Will," my mother pointed out again as she paced in front of my father, who was seated in his usual spot on the couch, sipping scotch.

"Well, why don't we ask her," my father said, suddenly aware of my presence.

"Ali, why are you so wet?" my mother asked.

"I walked home. I wanted to get wet," I said. "Anyway, I know a boarding school I can go to: Berkshire. I'll get in there."

"Are you sick?" She walked toward me and reached for my forehead. I yanked my head away.

"I'm not sick," I said. "Are you listening? I can go to Berkshire Boarding School. It's not that hard to get in there."

"Ali, please talk to us. What's going on with you?" She wiped her eyes with one of the tissues she carried in the left sleeve of her blouse. "Why would you allow yourself to get kicked out of one of the best schools in the country?"

"I didn't allow myself," I said. "It just happened."

"Alison, this is serious," my father said. "This is very serious."

I looked at him and felt a surge of unbearable guilt. His eyes, those disapproving eighty-year-old eyes, never failed to shame me. I began to cry.

"I know it's serious, Dad, and I'm sorry," I whispered. "I don't know how it happened. I didn't mean for this to happen, really, really I didn't."

"Well, it's no one's fault but yours. I don't understand what happened to you; you were such a great kid."

"Yeah, well, I don't know either," I said.

My father frowned sadly. My mother sighed and pressed her lips tightly together as if her heart was breaking for me, but I knew it wasn't. It was breaking for her. We all stood there in the hazy reddish glow of the library looking grimly at the beige carpet. After a few seconds of silence my father clicked the television on to watch the evening news, no longer acknowledging us.

"Go to your room, Alison," my mother said.

I obeyed and walked silently away.

What I remember best about the Berkshire Boarding School—where I did eventually end up in the aftermath of Spence—was our nightly showers. I can still see the bathroom. I can still see myself jamming the plastic wedge that read *Property of the Berkshire School* in bold black letters under the sliding wood door so that the bathroom was completely open and my music could be heard from the showers.

I can still hear the angry, maudlin noise blasting down the hall from my bedroom stereo as the less fitted doors shuddered in their frames. Minnie, my tall, curvy roommate, would appear wrapped in her monogrammed peach towel, her full chest protruding, begging the terry cloth tuck to unravel and the towel to fall. In one hand she'd hold her white basket of necessary Clinique beauty products, and in the other a pack of Camel Lights containing cigarettes, three joints, and occasionally a bag of cocaine. Aiden, my best friend at Berkshire, would lag behind her still dressed.

Though she rarely showered and didn't bother with perfume, Aiden managed to smell fine. It wasn't that she had an acute problem with being clean; it was just too big an effort for her to hassle with the whole process of showering. Gathering the shampoo, conditioner, soap, and razor; getting undressed; finding a clean towel; washing; getting out; drying off; brushing her hair. It was all too much, though occasionally we convinced her with a joint.

The Berkshire bathroom was a long beige-tiled room consisting of eight shower stalls, three toilets, and four ceramic sinks. The showers were at the end of the room, four on one side, four on the other, with a communal space in the center and a couple of plastic classroom chairs. On the far wall wooden hooks held robes or towels, and a vent above it allowed steam to escape.

Once inside the bathroom Minnie would light the joint, and we'd pass it around quickly while we had the showers to ourselves. Sometimes we'd argue or gossip; other times we'd smoke silently to the placating sound of water falling on tile as the steam humidified the pot and took it to the next level of potency. When we were sufficiently high, we'd peel our thighs from the plastic chairs and throw ourselves wildly around the stalls, wet and naked, to song after song that tumbled from our room down the hallway and into the bathroom. Eventually, we'd disperse into the various stalls and wash.

Soon other students would begin to file in for their evening showers. They'd shoot us looks of abhorrence and disgust, clasping their terrycloth robes tightly around their waists. They looked at us like our parents looked at us, like people look at squatters in ripped jean jackets, or junkies nodding out on park benches with transparent skin and perilously thin dogs. Naturally we had to fight back. We skipped around them, circled them like maypoles, sometimes tearing at the belts of their chastely tied robes, or ripping down their shower curtains, singing manically, smashing around—*down in a hole, feeling so low, I wish I could die!* We were like feral animals released into the world to display our undomesticated selves. It was the oldest and most disgusting story in the book, the self-loathing bully stepping on the weak.

Truthfully, it was frightening for me to watch those girls exist so contentedly. I couldn't stand the presence of their calm, comfortable selves, and I attacked them for living better than I knew how, attacked them for the seemingly effortless control they exuded over their everyday lives, when I barely had control over my next breath.

"Loosen up a little, Kay," Minnie yelled once, as Kay stood naked and frozen, her underdeveloped breasts perky, nipples dark and hard.

"Can't I shower in peace, Minnie?" Kay pled shyly.

"Here, Kay, have some of my cigarette. You really should smoke. It's very glamorous," I'd say, shoving it into her mouth.

Aiden, now wrapped in her ratty, unwashed towel, usually stood in the shadowy corner near the vent, chain-smoking and observing us with detached voyeurism.

"Don't you realize that not everyone wants to listen to your stupid music while they shower?" Jane Smothers piped up.

Jane was the head of the debate club, the council for the winter formal, and the field hockey team. She was the one overachiever in the pool of underachievers, and I am still not sure how she ended up at Berkshire.

"No, actually, Jane," I said snarkily, "we thought we were doing you all a favor."

"Fuck you," she snapped back.

At some point during most of our reckless adventures, Aiden would need escape and pull me aside with a whisper. I always did whatever she asked of me because the terror in her eyes was too sobering to ignore. She had many phobias: wet cement, cracks in the floor, British accents, alarm clocks—but she wasn't scared of scuba diving or roller coasters or dropping acid. It was the little things in life that scared Aiden.

Often at the end of the day, we would climb into sweatpants and bury ourselves under her plaid duvet cover, passing roaches clamped between the charred ends of tweezers back and forth. Smoke would trail up to the ceiling in a thick whirlwind, and we'd chat, dry-mouthed and sleepy, about whatever floated into our heads. Usually Aiden would nod out first, her eyelids fighting and then eventually giving into the tug of exhaustion. I loved watching Aiden sleep because it was the only time her face seemed to soften into itself, and often I'd stay awake guarding her, making sure nothing came between her and this rare peace.

I ran a lot at Berkshire. Away from teachers when they caught me smoking, across the campus if I was late for class, or around the track for mandatory sports. I ran up mountains as the arctic air stabbed my chest. I ran until my lungs ached, and I collapsed against the trunk of a snow-encrusted tree, my cheeks and fingers red and swollen from the glorious and biting night air. I ran to feel alive. Once I reached the top of the ice-blue peak, I'd lie on the hard, white layer of ground, my body slanting downward with the angle of the mountain, leafless trees towering over me, dividing the sky with their twisting branches like blue veins divide an expanse of flesh.

I also began to see more men at Berkshire. I had lost my virginity during a summer fling in Great Neck, Long Island, two summers ago. The boy had been sweet but dull. We'd done it in his father's waterbed and then he'd taken me to the train. I didn't see what the big deal was, but I did feel older, more mature, a little cooler.

I met Elliot formally the summer before I was sent to Berkshire though I'd observed and adored him from a distance for years at the Tuxedo Park pool. Back then, I was still a boney child in a navy Speedo and flip-flops, and Elliot was the club soccer captain, all bronze with hazy waterlogged eyes, a leather guitar strap hanging over his naked chest, and loose, untied Converse sneakers. He was older than I, a senior when I was a freshman. But suddenly, the summer before I arrived at Berkshire, he began to notice me. He flirted with me in a coy, manipulative manner, and I followed him around like a pathetic puppy. Sometimes we'd be out in a big group, and after ignoring me all night, he'd grab my waist and pull me behind a car, let me touch him, smell him. Sometimes he'd pull me into an empty street and kiss me ravenously. He'd have a quick feel of my still-developing breasts, and then he'd turn back to the group. I only saw Elliot on school breaks. Sometimes we'd have sex; sometimes we'd barely speak. Sometimes his behavior frightened me, and sometimes it turned me on.

During the school year, there were other men. A short snowboarder named Shane snuck through my dorm window in the middle of the night, cuddled with me, fucked me, and ignored me the next day. A bookish runner seduced me in the Berkshire storage room, yanked at my flesh like it was dough, and slapped me when he came with a strange sort of cooing sound. Then there was the effeminate dope addict who'd been such a gentlemen all night, bought me Midori Sours at Tunnel, held the door for me, paid for cabs, wrapped his cashmere scarf around my neck when it began to snow. It came as a real shock later that night when he began whispering—*You're so fucking hot. You're a fucking hot slut. A fucking hot slut bitch*—in my ear during sex.

Some nights at Berkshire my mother would call drunk two or three times, forgetting that she had already spoken with me. When I reminded her, she'd either brush it aside or attack me for lying, depending on her mood. She was always an erratic and temperamental presence, one I couldn't count on or vouch for, and I had learned not to depend on her for anything. My father, on the other hand, never called.

Our room at Berkshire was always flickering: ashtrays, clothes, and lava lamps haphazardly littering the floor. We stacked battered trunks with oversized gold latches below the window, and on top of them, Aiden's metallic silver JVC bookshelf stereo with a black-and-white sticker saying *If you want to fuck with the eagles, you better learn to fly.* My old Panasonic, which had been handed down from my half-sister's days at Taft, sat on the other trunk with a piece of loose-leaf paper taped to one side, a comically tragic Bukowski poem penciled across it. And *Dear Diary, I want to kill and you have to believe it's for more than selfish reasons,* a favorite quote from the cult movie of our time, *Heathers,* was stenciled on the other. The room smelled of beer, molding carpet, stale smoke, and occasionally incense, when we remembered to light a stick. Aiden's paisley rug defaced with cigarette burns appeared to melt into the wooden floorboards beneath it, and the blinds were always drawn firmly across the windows, taped at the bottom as if the most diminutive amount of light might painfully burn our skin. We spent our nights in circles on the bed or the floor, chugging vodka, consuming drugs, some depressing music always playing.

Aiden was always the worst—not to say that she consumed the most drugs because I think that was me—just that she was the most frightening on the scale of how far she'd gone over the edge. She had an acute and dangerous manic depression that had been handed down from generation to generation. Her grandmother had spent time in Silver Hill, a mental hospital in Connecticut where Edie Sedgwick had spent years getting shock treatments in the seventies, and her aunt, a prominent figure in New York high society, had thrown herself from her Fifth Avenue apartment in the middle of the night.

Although Aiden had inherited her grandmother's erratic beauty, depression and addiction were aging her, lining her face, yellowing her fingers, accentuating the heavy half-moons that already hung below her eyes. When she spoke, her thoughts seemed to crawl out from the far corners of her mind and rarely made sense. They were as ephemeral as the chains of smoke always leaking eerily from her lips, wafting past lamp after lamp and turning florescent before finally vanishing.

One evening late in November, Minnie sat on the corner of my bed, slipping a loose Marlboro from her shirt pocket as she did every few minutes. Aiden was lying on her bed where she had been all day, refusing to get up for class or meals.

"How the hell can you lie there for so long?" Minnie asked her, exhaling smoke across my face.

"Easy," Aiden said.

Squatting on soiled clothes that served as a cushion, I poured a baggie of white rocks onto a Sinéad O'Connor CD case and began to hum. Aiden leaned across me, propping her chin on her hand and grabbing a safety pin from her desk. I crushed the rocks with my school ID card. The more rocks in the baggie, the longer it took to crush, but the better the cocaine was. There was less filler—talcum powder, glucose powder, or baby's milk—cutting it, so I never complained. I crushed until I had a downy white mountain. I had always been a fan of the whole ritualistic aspect of drug use: the racking of the lines, the *tap tap* of the ID card against the CD or mirror or glass coffee table—whatever shiny, nonporous surface was around. The creation of the perfect white segments floating there like freedom, completely unattached to the rigors of the physical world.

Minnie jumped toward me as soon as she saw that the lines were completed, a straw already between her fingers and a half-foot from her nose. I stared out the window, wishing it would rain and waiting for the drawn-out sniff that would signal my turn. As I glanced in Aiden's direction, I noticed two long red scrapes on her wrist, a jagged line struggling upward through a maze of half-healed scars. The brassy smell of blood lingered in the room for a moment, and I quickly lit a stick of incense.

Aiden turned to face the wall.

I had guessed that Aiden cut herself for a while. I noticed small cuts here and there. In fits of rage, I too had hacked into my arm, scratched myself with a kitchen knife or a pushpin from my bulletin board. I would draw a little blood and recoil at the sight. But Aiden had taken it to a different level. For her each slice was a step closer to destroying the depression. Cutting was her key to salvation. We all had them, little Band-Aids we placed on the open wounds so we didn't have to look at them all the time. Clever tricks we'd perfected over the years to make life endurable. For me it was substance abuse, and for Aiden it was self-mutilation.

What happened next still puzzles me. I suppose the headmistress of Berkshire might have been reporting my misconduct to my parents without my knowledge, my constant demerits, my dropping grades, my reckless behavior. Or perhaps Berkshire had always been a temporary solution in the heads of my parents, a transient stop. They might have been planning to ship me away since my expulsion from Spence. Whatever the case, they arrived unexpectedly at Berkshire on February 2, 1994.

I could see my mother's face, sallow and anxious, through the wet windshield of the maroon station wagon parked outside my dorm. My father was reading the newspaper. The wipers were on full speed and whipped at the layers of water pouring across the car. Neither of them gave any indication of exiting the vehicle as if they were still undecided about whether to go through with it.

"My parents are here," I said to my friends, standing by the upstairs window.

"What are you talking about?" Minnie asked, pushing up beside me.

"See." I pointed to them.

"Shit!" Minnie screamed, laughing, jabbing me in the waist.

"Why?" asked Aiden from her bed.

"I don't know." There was a strange feeling in my belly.

The driver's side door opened and a brown plaid umbrella popped out, blew around with the wind for a minute and then opened. Now and then a twig snapped from the bare branches circling the road beyond the dorm building, the tennis courts, the track field in the distance. The class bell rang. I thought for a moment about running to class; there was a sense of safety in the thought of a chalk-smelling classroom with its blackboard and books and wooden desks filled with my peers. But instead I just stood there, watching them.

"Well, go find out," Aiden said.

"I guess I should," I said, but didn't move.

"Maybe your grandmother died or something," Minnie suggested, lighting a cigarette.

"She's already dead," I said, and walked out of the room.

I met my mother barefoot in the downstairs hallway, navy corduroys hanging low on my waist and a white V-neck T-shirt untucked and soiled.

"Ali," she said, shaking her umbrella onto the floor.

"What are you doing here?" I said in a perfunctory manner.

"We told you we were coming this weekend. We want you to come home. We have a meeting with a very good educational consultant."

Music blared down the hallway. My father appeared around the corner looking ashen, his trench coat drenched, droplets of water hanging off his ear lobes.

"I told you not to come. I'm not going. I have plans."

I turned and ran up the stairs at the end of the hallway, disappearing into a room. I heard my parents knocking on doors—*Is Alison Weaver here*—they asked over and over again. Finally they found me. An expression of mortification and disgust washed over their faces when they saw how I lived.

"Hello, girls," my mother said.

Minnie and Aiden, sitting on either side of me on the bed, said hello. I was crying.

"Ali, please don't cry. You'll see everyone on Monday."

Eventually I gave up and agreed to go with them. We walked out of the dorm and down the concrete pathway in the pouring rain, my father keeping a good distance behind my mother and me, purposely perhaps because he wanted to think about other things: roof painting, hinges that needed oiling, celery soup and melba toast. My mother shot ahead with purpose and determination, turning her head occasionally to make sure we were still following.

The landscape was gray as we drove, clouded over, wet with freezing rain. The rain whirled and surged, sputtered and poured this way and that, and high beams of white light gently ripped over our faces as cars flashed by. Night fell dim and cloudy as ashes. I curled up in the backseat and dozed off. I heard my father recline the passenger seat, place a United Airlines mask across his eyes, and begin to snore. My mother turned up the classical music on some AM station. It was Vivaldi's *Four Seasons*. "Winter" was playing. How fitting, I thought, and fell asleep to the tapping of ice pellets on the sunroof.

Two days later, I was at Cascade.

PART II

THEY WILL RUN OUR LIVES

Still, I did not despise my oddness, my deviations, those things which have made me, after all, me. I wanted to preserve those things.

—FREDERICK EXLEY

10.

A NEW FAMILY

I woke on February 4, 1994, on the upper bed of a rickety wooden bunk, my feet still numb, my eyes foggily picking out shapes in the darkness. Students were up, making beds, folding clothes, and buttoning shirts, the bathroom door in constant motion. I heard the sound of running water in the distance. If I hadn't known better, I might have thought it a sleepaway camp or a New England boarding school. I closed my eyes, and when I opened them again, a face was peering over my arm.

"Didn't you hear the wake-up bell?" she finally whispered.

"I guess not," I muttered.

"We have to be out of the dorms by 7:15 AM with all the dorm duties completed. I think you're on toilets, so I'd get up. Someone vomited last night." She picked a stray splinter from my bedpost and left.

I tossed the useless sheet from my body and climbed down the wooden ladder. It creaked. The uneven coarseness of ancient wood grated the soles of my feet. Outside wind pounded the building, loose branches were yanked off trees and ricocheted into the windows before falling onto the ground. As my eyes grew accustomed to the overcast morning light, I realized the girls around me were in the midst of much more than just typical morning routines. Some were on stepladders dusting shelves, others mopped the bathroom floor, or soft-scrubbed the showers. Stepping over a few crouched girls and soapy buckets, I made my way to a row of square compartments across from the showers, where I had been told to leave my toothbrush and toothpaste the previous night. The items looked lonely on the chipped plaster surface. The brush was new. It was red with glossy white bristles shaped like the jaws of a shark. My mother must have bought it for me. She probably threw it in with her regular order of Biotin shampoo and Basis soap. I could see her on the phone in the pantry—*Oh, and a toothbrush, one red toothbrush and a value-size tube of Crest Whitening toothpaste.* She was convinced my teeth had turned yellow from excessive smoking, and she wasn't going to pass up a chance to reverse the damage. When I returned home in two years, she hoped that not only would I be mentally up to par, but physically up to her standards as well.

"Alison, I'd do your dorm duties first," somebody said.

The face that had woken me earlier was now attached to a pudgy, pale-skinned body in a white fitted blouse that ballooned into large white puffs around her sturdy shoulders. She had straight yellow bangs, and as she spoke they swished back and forth across her forehead, falling into her close-set green eyes. Occasionally, she'd flick them into the air with a toss of her head or a blow of air from her mouth. She offered me a spray bottle of blue liquid cleaner, a circular brush with worn brown edges, and a neon green sponge.

"I'm Becca, by the way," she said. "Your Dorm Head."

I nodded silently, took the products from her hands, and turned into a free bathroom stall, which emanated the vile, unmistakable smell of urine. Near the toilet were a few pieces of balled paper that had missed the bowl, a rubber hair tie coiled tightly with dark black hair, and an empty Tampax wrapper. I brushed them aside, lifted the seat, and crouched on the floor, squirting at the yellow amoeba-shaped rings along the rim of the porcelain bowl until they were soaked in blue liquid. Then I spent the next twenty minutes scrubbing at them with the dilapidated bristles of my brush, all the while wanting to break into a rendition of "It's a Hard Knock Life for Us." Unable to resist the humor in the situation, I smirked to myself and wondered if any of the other girls were thinking the same thing. But when I tried to make eye contact with the one sweeping the floor in front of my stall, she immediately yanked her face in the opposite direction and returned to her sweeping with grim fervor.

It's difficult for me, now, to remember exactly what I was thinking during my first days at Cascade. It's so far from what I thought toward the end of my stay there, and the truth of it all is lost in between. Somewhere in the back of my mind I must have believed that this was another scare tactic; it seemed inconceivable that my parents would really send me away. I hoped that soon the gig would be up, and that I'd be allowed to return to my life. So I went through the motions: I ate when I was told to eat, cleaned when I was told to clean, and attended group therapy every day at 2 PM.

Rona, the bouncy redhead who'd led me through my initiation into Cascade life, reappeared later that morning. She found me in the dining room eating some artificial, neon-colored cereal.

"How was your first night?" she asked, animated and pert.

"I was freezing," I said.

"We can ask the counselors to get you another blanket," she assured me. "For now, though, we have the campus tour, and then you'll meet your new family."

The entrance to Cascade

"I don't want a new family," I shot back instinctively.

"I know it's hard; the first few days are always the hardest. It gets better though," she said, an expression of sympathy plastered across her "enlightened" face.

"Better than the strip search?" I drawled. "I can't *imagine*."

"You know what I mean," Rona said, looking agitated.

"I can't say that I do, Rona," I mocked.

"This is a good place, Alison. Just wait and see. I remember when I came, almost three years ago now." She tucked a stray curl behind her ear and looked off into the distance, nobly, as if she'd just been bestowed with the lifetime achievement award. "I'm leaving soon though—going to go to college."

"Yale?" I asked with bright insincerity.

"No," she said. "Puget Sound."

"Oh," I said. "Never heard of it."

Rona moved closer to me, placing an arm around my shoulder, nodding, stroking my hair a little bit, squeezing my arm with her stubby, red, nail-bitten fingers.

"I know you're angry," she said. "I know."

I shrugged her off, got up, and carried my tray to the garbage.

As we walked toward a long ranch-style building, Rona listed rules and pointed out smaller buildings, introducing them as we passed. She spoke of Time Limits, explaining that I couldn't spend more than ten minutes a day with any student who had been at the school for six months or less, thirty minutes with any student who had been at the school for six months to a year, and so on. She spoke of Sugar Limits, Juice Limits, and Shower Limits. She spoke of Privileges, said sharing anything was illegal unless I had been awarded the Sharing Privilege, for which a proposal must be written and approved by a board. Dating was illegal, as was kissing or hand-holding or cuddling with the opposite sex. Swearing, spitting, holes in clothes, untucked shirts—all illegal. Pens must be carried at all times, hair must be neat and combed, shoelaces must be tied, no running inside, no yelling.

When Rona finished, she dropped me off in a big beige room at the end of a long corridor in the building. Monstrous wooden sofas piled with checkered velour pillows lined the walls. Edges were sharp and protruding, windows thin and plain, rattling in their frames as gusts of wind beat the walls in unremitting fury. There were no curtains, no color anywhere. Ten or twelve students around my age sat on couches or in despondent little groups across the floor.

"This is your new Family," Rona explained. "Everyone here has arrived some time this week. Actually, I think you're the last one to join the group." She turned to the teenagers in the room and gestured to me.

"Everyone, this is Alison. She's a new arrival, so please make her feel welcome." She patted me on the back and left with a final perky grin.

Scrutinizing the wan faces staring up at me, I backed myself into a corner and slid to the ground. Some shivered, while others drew designs on dry skin. One girl picked at her split ends meticulously. Another boy held his middle figure in the air, presumably directed at Rona or me. Three girls giggled to each other in the corner of the room like normal schoolgirls gossiping, though a second glance caught something wild and manic in their laughter and movements. A younger girl sat silently with her legs folded under her on the corner of the sofa, staring vacantly into the room with a medicated expression, softly cradling her face. It was almost as if her mind had checked out and it was only out of habit that she moved and spoke with the rest of us. A desperation like none I had ever felt lingered in the air of that room, ran through every jagged splinter, every chip of hardened paint.

"Where are you from?" the boy next to me asked, glancing down from the armchair above me.

"New York City."

"Guess it's easy to get in trouble there." He pursed his lips upward and wrinkled his nose downward until they merged into one another.

"No easier than anywhere else," I said passively.

"I'm from Portland. Name's Sam," he said, thrusting a hand in my direction.

"Oregon?" I asked.

"Yeah, Portland, Oregon." He sat up proudly.

"I don't know it well."

"Deidra says that I was conceived in New York," he said.

"Who's Deidra?" I asked.

"My mother," he said.

"Why don't you just call her Mom?"

"She doesn't want me to." He squeezed his lips to his nose again.

"Oh," I said.

"Want me to leave you alone?" he asked.

"No, I mean, you're not bothering me. I just don't feel very well."

"The first few days are real hard," he said, and stopped talking to me anyway.

Sam was my first friend at Cascade. He was a computer genius, with an IQ of well over 175. But his intellect terrified him, so he bent his enormous will towards becoming a professional skater. By age ten he was signed with a label and was about to *go pro,* as he called it, when Deidra sent him away. He was an awkward boy, thin-boned and floppy, with a nose widening flatly across his face and plump puckered lips that squeezed into it whenever he was nervous. He projected an endearing stupidity that offered him a layer of protection from the reality he never wanted to face. You could tell he was a skater by his attire: the thick, padded Skechers sneakers; the baggy corduroys; and the orange T-shirt that still retained the outline of the skater brand logo the school had removed. Cascade tried to wipe out your identity, but there was always a bit that remained leaking through wherever possible.

Family Time lasted for thirty minutes. I was introduced to the other children and assigned an individual counselor with whom I was to meet once a week for two hours. When the meeting broke up, I filed silently behind the others toward the next destination, which according to the schedule Rona had given me, was first period—U.S. History. I spent the next hour being lectured about the War of 1812 and James Madison's second inauguration while feigning diligent note-taking. The kids I recognized from Family Time were not paying attention at all. The boy in front of me was tagging across the white loose-leaf paper inside his black binder, and the girl next to him was drawing bubbly flowers and purple

VW buses along the margins of her yellow notebook, occasionally jot-
ting down something the white-haired instructor said. Sam, who had
plopped himself down beside me, was sketching a portrait of a much
older boy in the front row.

Many of the kids in the classroom seemed half-dead, their eyes
swollen and shadowy, glossed over, beaten down with commands and
punishments. Even the ones who posed questions exuberantly didn't
seem real but rather like second-rate actors forcing their parts upon a
disenchanted audience. But what I found most disturbing about the
school were the peculiar counselors in their clogs and brightly colored sweat suits, whose eyes were wet with compassion and empathy,
whose hands wanted to pet and coddle, and who all spoke in the same
slow, peaceful drawl. These counselors would chart our progress, deter-
mine our prognosis, and draw up an appropriate treatment. They
would decide our privileges and our punishments. They would run our
lives.

11.

FORUMS

Cascade was set up in groups called Families, instead of grades. A Family was a group of about eighteen to twenty kids who had arrived at the school around the same time, usually within one month of each other. The longer your Family was at Cascade, the higher up in the program it moved and the more privileges and respect you earned—assuming you continued to follow the rules. Cascade consisted of four Schools: Beginning School, Middle School, Upper School, and Leadership. There were usually two or three Families in each School. If a member of a Family that was due to move up a School was not behaving—that is, following the rules, working on themselves therapeutically, and contributing to "our" community—then they were demoted to a lower Family. The standard stay at Cascade was two to three years, depending on your level of growth and your commitment to the school.

A student's years were marked along the way by eight therapeutic workshops: The Truth, The Youth, The Friends, The Sisters/Brothers, The Heroes, The Imagine, The I and Me, and The Symposium. Each Workshop was a weekend spent in a renovated barn called the Millhouse, participating in "therapeutic exercises" of one sort or another: pillow pounding, mask wearing, towel pulling, visualizations, dancing, yelling, and crying, always crying. To complete the program and earn a Cascade diploma, you had to successfully "conquer" all eight Workshops and have graduated into Leadership School. Of course, some people left as soon as they turned eighteen without finishing the program, but most stayed because by that point they were usually steadfastly dedicated to Cascade and its teachings.

We slept in six separate dormitory cabins, three for the boys and three for the girls, each divided into six sections that were separated by wooden partitions. The dorms were decorated as if we were children dying of Ewing's sarcoma or leukemia at the Ronald McDonald House. Stuffed animals were strewn across the beds; colorful paintings of sunshine and flowers and messages of hope and perseverance were taped to the walls. The bathrooms consisted of three bare and cavernous showers, a wall of cubbyholes, where we kept permitted toiletries, and three toilets and sinks kept clean by our daily scrubbing.

Tuition at Cascade was a staggering hundred thousand dollars per year. Because Cascade was an accredited high school as well as a therapeutic rehabilitation program, they somehow got away with charging this ludicrous sum. But the education they provided for us was poor compared to the education I had received at Spence and even Berkshire. Most children who were sent to Cascade came from very affluent families, but some were from the middle class—children whose parents and grandparents had mortgaged their houses, scraped together every last penny, destroyed college savings or retirement funds to save their wayward children.

Cascade's primary purpose was to promote emotional growth for its students, which accounts for the excessive amount of therapy we were forced to endure. Group therapy sessions were termed Forums—three-hour sessions of twelve or thirteen students and one counselor. They were loud, unpredictable, explosive events that frightened even the toughest student. Kids cried about various issues, screamed across the room at their loathed adversaries, or bellowed at the floor while Running Anger. "Running Anger" was a Cascade term used to describe a student screaming at the floor while imagining the person who had hurt him or her. Legs were positioned in a V shape as the student rocked back and forth, their rage and grief growing stronger with every forward motion until the torso was slamming forward, head thrusting up and down in the center of the V, faces flaming, bursting with angry air like an overfilled helium balloon.

Few things in life will ever surmount the terror I felt inside the white-washed walls of those forum rooms during my first few sessions. Anything could happen, and often did, each event more volatile than the next: bloody noses, popped blood vessels, vomit, chairs being thrown. Every time I was sure I had heard the worst story, another blew it away. Fathers and uncles raping children, mothers committing suicide, brothers dying of AIDS. Molestations, beatings with scalding hot water and Brillo pads, brooms, studded belts. The cruelty of humanity multiplied before me.

Quickly, I grew afraid and ashamed of the smallness of my own stories. Silent secrets are like air; you can't see them or define them or point at them and say *this hurt me and that is why I am the way I am*. You can't Run Anger about hints of melancholy, or the tears you saw in your mother's eyes. And sitting in those Forums day after day, my complaints devolved into petty and trivial nothings. Why did I hate my parents? I couldn't even remember.

Forums were scary enough to observe, but once I began to be Indicted, they became unbearable. After my first month or two of sitting silently,

The building where all our Forums took place

hands clasped in my lap, doing my best to remain invisible, I was Indicted by Rona Crane. The room was silent. No one felt like talking that day. The sky was a blanket of neon blue, electric and charged with energy; it beckoned us into the mountains, begged us to leave the musty classroom and enjoy the first mild April day. The counselor noticed and shut the shades. Rona stood up and marched heroically across the room to take the seat facing me. This was part of the Indictment procedure: You had to be directly across from the person to whom you planned to speak. When I realized she was coming for me, I sank into myself.

"Alison, I'm worried that you're just sliding through the program without really working on yourself. I don't trust you, and I think you break the rules, but no one notices. You never talk in Forums, and you never indict. I think you need to start working on yourself and make a contribution to our world here." Rona seemed angry, although she modified her expression with a bit of tenderness.

I scanned the room trying to gauge the reactions of the others. Older students looked deeply concerned about me, stroking their chins or biting their lower lips and nodding. The newer ones picked their split ends or stared blankly into space, relieved the attention was now on me.

"I follow the rules," I said.

Ernest Hoffman, the counselor running the Forum, spoke up. "Alison, I've run three Forums that you have been a part of over the last few months, and in not one have you uttered a word. I don't think the point is whether or not you follow the rules, even though I will be giving you level three dishes for the next week because I trust that Rona knows what she's talking about. Let's leave the surface. The point is that you need to put forth effort to help yourself. Why don't you do some work today?"

Ernest was short, with tight muscular legs, a broad, overdeveloped chest, and a thick mustache. He wore loose tank tops and white high-top sneakers and always smelled of mint and cigars. You could find him on the basketball court most days, whooping and hollering over some phenomenal shot or mistaken foot fault. He was an ex-addict—a heroin junkie for most of his twenties—and one of the founders of Cascade. The odd thing about Cascade was that it was founded and run by a group of ex-addicts, not one of whom had a psychology degree of any sort. No PhDs, no MDs, no therapists or social workers anywhere. All the people who worked there were called "counselors," a rather obscure title.

After Ernest spoke, the entire group stared at me, waiting, I imagine, for me to break into the saga of my life. But my mind was blank. I had nothing to say, nothing that compared to these people's stories. At fifteen, I believed that my depression, my insanity or lunacy or stupidity, was my own. No one had done anything to me.

"You think you're better than we are, I know it," Becca spoke up.

"I think you're a real bitch, Alison. I tried to talk to you in math class the other day, and you were so cold, like *icy* cold. You do think you're better than us," another girl insisted.

"Is that the problem? You think you're too good for this place?" Ernest said.

"No, I just don't think I belong here. I don't have anything to talk about," I said.

"Oh, you don't belong here? You're better than us? You're just fine? You don't need any help?" Ernest prodded.

"I don't need this," I insisted quietly.

"You've given me that bullshit since day one!" Rona screamed across the room. "When I tried to comfort you, you pushed me away, pretended you didn't need anyone, pretended you didn't need Cascade. You do think you're better than us, I know you do. I've tried to be your friend, but you think you're too cool or something, and it pisses me off!"

"Why are you here, Alison?" Ernest asked calmly, resting his chin in his palm as if his now-comforting tone would trick me into opening up.

"Because I was doing drugs, and I didn't really get along with my parents," I said.

"So you were doing drugs because you felt content and happy in your life? Is that why?"

"No, I am not saying I was happy," I said, glaring at him. "I know things were fucked up, but I didn't need to be sent away to this freak school. I could have gotten help at home. I shouldn't have been ripped out of my life and sent across the country."

"Clearly you couldn't be helped at home, Alison. Haven't you been in therapy since you were five?"

"Yes, but that was different!"

"Oh, really," he said with clear disbelief.

The Forum went on in this manner for a few more minutes, but I didn't talk again. I was too tired to fight. I just sat there, judging them, as their mouths moved and big clouds of noise drifted out. I just sat there believing myself better than them—just like they were telling me. After all, that's what I'd been taught my entire life. Sure, I could have faked it. I could have pretended I cared. I could have manipulated them into liking me because I could make anyone like me, but I didn't have the energy to bother.

I ate dinner alone that evening. The dining room seemed blurred. The noises and the people magnified and shrunk over and over again as Pachelbel's *Canon* played redundantly on the stereo. I wasn't hungry but attending dinner was a rule so I sat obediently at the wooden table and played with my food, flattening my scoop of mashed potatoes with a fork until it looked like a cloud of highways leading to the end of the earth. I would have given anything to chug down a bottle of vodka or slide my blue and red striped straw across a plastic CD cover. I longed for that vile cocaine drip to line my throat. I longed for detachment, complete and utter numbness. Everything I had grown up believing was coming into question. I was being asked to open up and talk about unspoken family secrets, things that I'd barely even allowed myself to think, let alone speak out loud. After all, I still held my mother's belief that acknowledgment of anything unusual was somehow shameful: depression, alcoholism, divorce, adultery, homosexuality—these things were simply not discussed.

I understand now why I fell for Cascade. Why I eventually gave up fighting and let it sweep me away with its preaching and its cultlike rules. Cascade offered me a place where I was free to be open about my life, and it removed me from a world in which I had so clearly failed. I believe that's why most of us fell for the place; it became our sanctuary, our safe haven. The counselors beckoned us like universal fairy godmothers; they rubbed our backs, stroked our hair, and cuddled us to smooth jazz. They listened to complaints about our parents and shook their heads at the behavior we described. *Come relax for a few years. It's much nicer here than out there,* they'd say. We believed them. *I* believed them.

I walked alone to my dorm that night. The room was empty, the others still eating. I wasn't supposed to be in the dorm by myself; privacy was another of Cascade's many Privileges to be earned. I was supposed to ask an older student to come with me, but I didn't care. It felt great to be alone. I wanted to disappear quickly and quietly, pack my bags and take a bus far, far away, but I knew that wasn't possible. Instead, I ran a hot shower. The steam was white and still and it swelled like safety around me. Resting my back against the mildewed tile I slid inside it and cried.

"Alison," I heard someone call. "Alison Weaver."

There was banging on the bathroom stall.

"Yes," I called back.

"Alison, you're not allowed in the dorms alone, and you're supposed to be on Dishes right now—remember?" It was Rona.

"Oh, I—"

"Get dressed and get up to the kitchen!" Rona yelled. Other people were with her. I could hear them all huffing to each other about my behavior.

I turned the faucets off, dried, and dressed rapidly. Running through the cold, dark night to the dining hall, my heart pounding, I could see my breath trailing in thick grayish puffs like the exhaust of airplanes leaving their ghostly shadows in the sky. It had begun to snow lightly. Large flakes fell on my loose wet hair as a calming whiteness spread across the campus.

Dishes had already started when I entered the kitchen. The building was silent except for the whirring of various machines and running water. One student in navy sweatpants and a baggy, soiled T-shirt paced back and forth with a clipboard, overseeing all the jobs. I spent the next four hours scrubbing the floor of the back kitchen pantry and ten large rubber mats caked with food. Occasionally, I'd stop cleaning and watch the snow that had begun falling in fat streaks from the sky, looking almost blue, as the wind tapped against it gently, letting it hover in the air for a moment, before picking up speed and thrashing it into the building.

12

A SPLIT CONTRACT

Two months after my arrival at Cascade, I was allowed my first ten-minute, counselor-accompanied phone call with my parents.

"Hi, sweetheart," my mother said. "How are you? I've missed you."

"Will you take me home?" I asked.

"No, dear. This is a final decision. Your father and I didn't know what else to do with you. Can we please talk about something nice? This is the only time we have for another two months," she replied.

I heard her take a sip of wine and was comforted by the sound of her lips against the glass and the slow, determined swallow. It's strange what we learn to find comfort in.

"Mom, I know you're mad at me. I know I've been difficult, horrible. But please, I'm your *daughter*," I said.

"Ali, you are my daughter, and I love you very much and only want the best for you. Cascade is going to help you. You'll be grateful in the end, sweetheart. You'll be a better person for it."

A better person for it? What was I supposed to say to that? She talked on and on, a spate of vacuous information: weather forecasts, stories about my dog, our neighbors, our cook, upcoming trips my father and she were taking. Half listening, I began to picture her at home at her desk, a brown wooden desk with white circular stains spotting the shiny finish, catalogues stacked in piles to the left and the right, invitations to benefits, charities asking for money, all scattered across the top. My mother was always sitting at that desk, white wine in hand, sipping, swallowing. The color of the wine harmonized nicely with her, the yellow-blond of her dyed hair, the yellowy-red of her skin and nails. My mother seemed most complete behind that desk in the study, where she sat for hours every day, doing nothing at all but smoking cigarettes, sipping wine, and leafing through papers.

When I think back to my childhood, I don't remember my mother as a true-to-life figure, but rather as a flat image—an icon symbolic of a woman who I believed to be my mother. My memories of her are in the form of still images: sepia prints, not colorful home videos. They don't play in my head like a reel of film but appear rather as frozen frames, and the most familiar one is of my mother behind that desk.

I never knew who she was as a child. I don't really know now. So much of her was hidden from me: who her real father was, her stepfather's death, and her mother—my grandmother—who visited once every few years with a small suitcase containing grape and cherry candy. My grandmother hated the idea of being a grandparent and made me call her Roz, a nickname "the girls at the home" had given her. She drank bourbon, usually sloshing it all over the floor as she lifted the glass to her lips—lips of an old woman with unevenly distributed cherry lipstick. The night Roz died my parents were out for dinner. I was eight. The lady from the nursing home called around seven: *Rosemary Fine passed away at 8:12 PM.* I collected the few pictures of her that were dispersed around the house: one of me on her lap, one of her hitting a golf ball, and one of her as a teenager dressed for prom. I set them up on my night table and bunched up tissues, placing them in between the frames and on the floor. When my mother came home later, I assumed she would check on me, see the pictures, and deduce the death herself. I guess she must have, though we never spoke about it until years later. If there was a funeral, I didn't go.

My mother and I couldn't communicate loss. She never told me when my goldfish died. One afternoon, I found the bowl cleaned and sitting empty except for a bag of hexagonal orange pebbles; a week later a new fish appeared. She didn't tell me when she put either of my dogs to sleep, nor did she tell me when my father died.

As I sat in my counselor's office picturing this woman on the far end of the line, I realized that I didn't feel any love for her, not at that moment. She was "my mother," and beyond that I didn't know.

"I have nothing else to say," I said to her.

My counselor, Tina, disappointedly shook her head. She was dressed in loose paisley overalls and hemp sandals. She always looked as if she were on her way to a Dead concert or a peace march in San Francisco, and I had trouble taking anything she said very seriously. Years later, I'd find out that she was smoking the pot confiscated from new arrivals with a few of her favorite male students.

Tina took the phone from me with a disapproving glance, and tucked it between her chin and shoulder. "Yes, Mrs. Weaver, yes, of course. It just takes time. All right then," she said soothingly, the line of her back blocking the phone from my sight.

Her sandals left dried mud on the carpet as she spun and swiveled her chair this way and that, finally hanging up the phone, sliding toward me, and draping one arm around my shoulders with a sigh. I knew exactly what she thought of me, I could see it in her eyes—a spoiled, ungrateful brat. The sigh was disgust she couldn't express. The one-armed hug was her half-hearted attempt to mask it.

"Alison, your parents are hurting too. You're not the only one in this world, you know? You have to give a little," Tina said in a tone a nanny might use to spell out inane instructions to a dim-witted child.

"I know at home you were their little princess, and you got whatever you wanted, but not here, okay?" she said, her arm still around me.

"You don't know anything about my parents or my life," I said. "I don't really want to talk, anyway. I have homework to do." Rising, I yanked my backpack off the ground.

"Sit back down, Alison. Our session isn't up yet. Just because you end a phone call early doesn't mean you get to leave early," she said. "If you don't start working on yourself, you won't grow here, and you won't graduate."

"I am *not* staying here. You think I'm going to be here for two years?—that's a joke," I snarled at her. She watched contemptuously as

I slammed the glass door behind me, edges jiggling within the wood frame.

Communication with the outside world was extremely limited at Cascade. We were allowed to correspond only with our immediate family members, and even then only under counselor supervision. The authorities encouraged letter writing, though all letters were censored before they were mailed. I wrote my father for the first time in early April. I have no recollection of what I said, but he responded from Vail, Colorado, on April 22:

Dear Alison,

Thanks for your letter of April 10. Hey—what's up? I like that line. I hope your bronchial pneumonia (not ammonia, which is what you clean hair brushes with) has gone away. What do you tell them in the group therapy or is that confidential? We miss you a lot out here in Vail. There is no one to get the first chair with me. The good news is that I have skied nine days and only fallen once and I bought a new pair of ski boots.

In New York the apartment elevator operators and doormen may go on strike. The last time that happened, Tony Quintos ran the building and we had to operate the elevators. See you next month. We love you.

Dad

I suppose my father didn't want to think about the frightening girl he had sent away a few months before. He didn't want to remember the scornful disgust he felt as he watched his youngest daughter parade past in torn, hip-hugging jeans, with bleached white hair and dark eyeliner. Didn't want to remember the dark music that pounded the walls of her bedroom, or the images of underground worlds that she'd pinned on her bulletin board, or the smells of smoke and incense that leaked into the hallway through the cracks in her locked door.

Out of my own desperation to hold onto something solid and meaningful from my old life, I had relapsed into complete adoration of my father and once again erected him as my hero. I believed him to be fault-less and directed all wrath and fury toward my mother.

Later that evening, I sat at a small dining room table with some members of my new Family and an older student who sat with us out of obliga-tion. Her face hovered above a plate as she sawed adamantly at a slab of fatty meat with a flimsy plastic knife. All utensils at Cascade were plastic. We smiled to ourselves—at this point, "we" were still a *we*, with the counselors and older students firmly designated as *they*—content for that second to bond over the ridiculousness of her behavior. After wolfing down what would hardly be considered four bite-sized pieces of meat, she rose and returned to the counter for more.

The meals at Cascade consisted of poorly cooked fast food ranging from mashed potatoes and jaundice-colored macaroni and cheese to soggy pizza, hamburgers, or chicken nuggets. The dining room was large, with tall, church-like ceilings of white stucco separated by splintering beams, all supported by authentic wooden posts. A shiny gray floor held hundreds of wooden tables and chairs, and a bearded man with the longest hair I'd ever seen buffed the floor every night with a huge metal contraption aptly called The Buffing Machine. It had handlebars and foot pedals, and he'd ride it around the dining room with great convic-tion, determined to cover every square inch of floor. On the walls of the room were gold-plated letters about three feet tall forming words: *In-tegrity, Hope, Acceptance, Forgiveness.*

"I've never seen anyone eat like that, have you?" I asked Sam while the older student was refilling her plate.

"Never, and she's so skinny," he said.

"She's obviously a buli," said Jinny.

Jinny was only thirteen, one of the Cascade babies. There were ten students there who were only thirteen and three who were twelve. Jinny dressed in loose black sweaters, tight stonewashed jeans, and gray high-top Converses. Her eyes were a liquid hazel that in certain lights looked as yel-low as the eyes of an owl. She was subdued and elusive, always speaking from behind her veil of thin, dark hair like the voice of Oz from behind the curtain.

"A what?" Sam said.

"A bulimic. She'll throw it up," Jinny whispered.

"Yeah, my cousin was one of those. She puked so much that the toilet screws rotted and the seat became loose," a voice from the other end piped up.

Sitting next to Sam was Jake Forester III, heir to a multimillion-dollar oil fortune. He had arrived the day before me with two trunks filled with Everclear and moonshine, thinking Cascade was just another boarding school. He was also the only student in the history of Cascade to successfully sneak an entire ounce of marijuana onto campus, despite the strip search. How he managed it one can only speculate. Jake was tall for fifteen, though lanky, with cotton-soft hair and powdery blue eyes. During his first summer at Cascade he refused to wear shorts, even on the warmest days, because he was tormented at his last boarding school for having "chicken legs." Bright eyed and bubbly, he possessed a child's abiding sweetness twisted with a cocky confidence that was often threatened when one of his famous escape plans or far-out theories crumbled.

Next to Jake was a pretty blond girl named Crissy McManister from Orange County, California. She giggled whenever Jake spoke, brushing against him with an ever-so-subtle glide, at which point he'd laugh his nasally laugh simply because Crissy was laughing, though nothing anyone had said could have been interpreted as remotely funny. Crissy's best friend, a tough Korean girl named Lara Brown, sat on the other side of her. Rumor had it she'd spent the last several years running with a dangerous Asian gang in Los Angeles and actually sliced a girl's cheek for kissing her boyfriend. Across from Lara was Mike, a shy boy from West Virginia, who stared at the ground when he spoke, refusing to make eye contact with anyone.

"I'm planning a split. You guys want in?" Jake asked, smiling and rubbing his palms together.

"Yes!" I said. "Please get us out of here."

"Yeah," Sam said. "How do we do it?"

"I got it all figured out. Meet at the cattle guard tonight at 4 AM," Jake said, nodding.

Plans to escape were drawn up daily at Cascade, some just talk, others failed attempts. Sometimes they were elaborate, complete with outside communication, covert phone calls, and friends who owned helicopters or could get fake passports, and others were basic—crawl through the window and run. They gave us an intangible hope, something most of us desperately needed during our first few months at Cascade. Only two people had ever managed a successful escape; they journeyed over the mountains with stolen camp equipment and were never found. As for me, I made it three steps beyond the sliding glass door that led out onto the deck of the dormitory, and after spending the next two months digging ditches and scrubbing grills, I never bothered again.

I quickly came to feel that most of the students at Cascade were much like the students I had roomed with at Berkshire. They didn't seem any crazier or any more unproductive. Of course, some were severely depressed, attempting half-hearted feats of suicide from time to time, but the really crazy ones didn't last long. They were usually sent to mental hospitals, lock-ups, or rigorous wilderness programs, three places we were often threatened with and feared inexplicably.

Cascade students, including Jake (at right)

During my stay, a large red helicopter landed on the campus a few times, usually during the class directly after lunch. I'm not sure what it was about this time of the day, what exactly drove people to self-destruction or self-medication. We'd hear the obnoxious *chop-chop-chop* of the propeller from our classroom windows and turn to watch the ducks on the pond flap dramatically about, creating storms of white feathers and water. Unable to focus on the lesson at hand, our minds would begin to imagine what caused the helicopter visit and whispers would begin. Finally, the teacher, realizing that continuing the lesson was futile, would let us out. It had become a sort of Cascade tradition: one hundred fifty students huddled in groups on the field exchanging the rumors that had already started. In a sense we were there to pay our respects to the student who had actually attempted what most of us never had the courage to do. A white stretcher would appear, carried by two hefty nurses, and we'd all hold our breath and wait to see who lay on top.

The first time I saw the helicopter it came for a girl named named Dianna D'Alaila. Short and round with white skin, tar-black hair, and natural strawberry-red lips, she snuck into the dorm after lunch, opened her mouth and proceeded to pour half a bottle of Windex down her throat, convinced this would kill her. She left in the helicopter, had her stomach pumped, and was sent to a lock-up facility two days later, deemed by Cascade to be "a danger to herself and in need of twenty-four-hour surveillance." The second helicopter visit was for a short blond boy named Bill who thought he could get drunk on Listerine. He arrived at breakfast stumbling from home fries to pancakes until his body, with a discordant noise, projectile-vomited across the dining room floor. He returned to Cascade after three short days, embarrassed and condemned to Level Three Dishes. A thin, freckled girl resembling Olive Oyl spent weeks out in the mountains picking grapes. She bottled them in water bottles and kept them hidden under her bed, sold on the idea that she could make red wine by fermenting them. After a few months, the grapes did indeed liquefy, and she proceeded to chug bottle after sour bottle. She was another who never returned.

Sam got taken away in a helicopter too. He spent the evening huffing glue inside a bathroom stall and then thought it would be fun to skate off the balcony rim without an actual skateboard. Sam could slide across anything if it was wet enough. It had just rained; the wood was soaked through, slippery enough to slide across if one had sufficient momentum, but Sam fell, cut his lip, and was knocked unconscious. He spent two weeks in the hospital before returning to us, luckily in one piece.

I loved it when the helicopter came. I think we all did. It added an element of excitement to our monotonous days. And we never for a moment thought anything serious had happened because a successful suicide was almost impossible at Cascade. Chugging detergent, Comet, Palmolive, or Listerine—these were just statements of exasperation, little *fuck you*'s to the Cascade system, nothing a stomach pump and a few days in the hospital couldn't fix.

13

THE TRUTH

After the initial shock of my new Cascade life subsided, a strange thing happened. I began to find escape and solace in the land. The school was set in a valley next to the Cascade mountain range dividing Oregon from California, and the blistering sunsets were like none I'd ever seen. The sun's rose-orange rays speared in and out of the peaks with the precision of a calligrapher's pen, spraying color past our mountains and deep into the distant sky above Oregon. I'd often imagine the same sun setting across the concrete building tops of New York City and find enormous comfort in the fact that I was still somehow connected to my old life—even if only by an orange ball of heat 150 million kilometers away.

The woods of Cascade were an endless thickness of green and brown that ran in one dark, fluent blaze over boundless mountains and sang fitfully with the clatter of nature. On the weekends, we were occasionally taken out on hikes through these woods. We'd walk for hours along the paths of half-exposed roots and faded boot prints lined by soaring pines and Douglas firs. I'd sweat and pant, and for the first few months my lungs ached, but I never stopped. I loved the taste of my salty sweat mixed with the trail's kicked up dirt, beading over my lip; I loved the fact that for those few hours I wasn't a marionette on the strings of Cascade. Hiking was the only time I felt like a real person.

* * *

But the day I entered my first Workshop I was unable to find comfort in those woods. The noises of the wild had ceased completely and the streaks of green forest lining the path to the Millhouse were marked by an eerie, anxious silence and an early morning fog so heavy and dense it seemed capable of suffocating us. The sky had clouded over, and it looked as if it might storm, but there was no wind, nothing moving. Ducks glided across the murky surface of the pond, occasionally diving under for a fish or a soggy piece of bread. I followed my Family along the path, their arms swinging in sync above the gravel, some hands tense and fisted, others seemingly relaxed. Jinny and Lara walked with linked arms, Jake had a casual though determined rhythm to his strut, and Sam, as usual, lagged behind, shuffling along as if he had just been injected with Thorazine. Behind the Millhouse building white fog began to settle around the mountains like the sheets of a Halloween ghost. On that morning it felt like we lived above the clouds.

The Truth:

> Inside the barn, we sit in a horseshoe as Karen Carpenter sings in the background.

> A woman named Marlene enters the room. She's pudgy, with dark, curly tufts of hair. Her voice is threatening, imperious. Her eyes burn my face like acid.

> "Welcome to your first Workshop! You will only get as much out of it as you are willing to put in. This is a pendulum, your first of the Cascade tools," she says and points to a large white pad with a black pendulum sketched in the center, one side labeled sorrow and the other joy.

> She continues: "If you only allow the pendulum to swing a little bit on one side, it will only ever swing a little bit on the other side. Agreed?"

> The group nods.

I am terrified. I know where this is going. She says we must embrace sorrow, so someday we can embrace joy. She says we must peel open the layers of pain and delve deeply into them before we can ever experience joy. I cringe. I can feel my heart tremble in my throat, and I have to sit on my hands to stop them from shaking. I don't know what my sorrow looks like or feels like, but I have a sneaking suspicion it is stronger than me.

The first exercise is called the Circle of Shame. We are told to go clockwise, one after the other, and admit everything we have ever done that we feel ashamed about. It takes hours and hours. Marlene doesn't accept what we say. She says we are hiding. We are. My body is numb. I am hungry. She yells at Jinny.

"When did you lose your virginity? Did you know the boy? I'm speaking to you, Jinny! Answer my question!"

Jinny cries and cries and rocks a little. Her face is in her hands, and her thin brown hair falls messily in every direction. She admits she was ten. She admits it was her uncle. She describes a pale pink room with clouds painted on the ceiling and a white bed with pink and yellow daisy shams. She says she stared at a watercolor of Little Miss Muffet painted by her grandmother the whole time. Marlene moves close to her and rubs her back. Karen Carpenter begins singing again.

Marlene turns to Jake, and he says he wants to pass. She says there is no passing. He says he has nothing to discuss and crosses his legs. He is wearing soiled jeans and has a blond bowl cut. He is fourteen years old and thinks he can manipulate Cascade the way he did the other eight boarding schools he had attended since he was nine. She orders him to leave the circle, and he walks across the room with flushing cheeks.

"I don't have time for this bullshit!" Marlene shrieks. "I am here to help you, but I need cooperation. I need honesty! I want you to peel back your flesh and show me the ugliest part of your soul. We have to break you down in order to build you back up!"

Marlene dissects us all. She demands that we all have dirty little secrets. Many do, and confessions flood the room. Boys confess to rubbing ice on their genitals, forcing their dogs to lick their testicles, tying cats to trees, torturing younger children. One raped a girl, one punched his mom, one molested a little girl. Almost everyone admits to having been beaten or molested by a family member. A few girls prostituted themselves for drugs, others attempted suicide or slept with their sisters' husbands. For a moment I consider making up a traumatic incident. At Cascade it seems that the more severe the abuse, the more gruesome the event, the more embarrassing the confession, the better off you are.

It is late when Marlene reaches me, and there are still three more students to go. I want to tell her this is unfair. I don't want to tell all these strangers my secrets, my shame. I want to scream at her for yelling at Jinny, but I am silent, frozen with fear.

"Well, Miss Weaver?" she says.

I shrug my shoulders, and tears fall from my eyes, so many that I can't catch them all with my tongue, but I try. Inside my mouth, I can taste their salt. I don't speak. Everyone is hungry because we haven't eaten since lunch ten hours ago, and they look at me angrily. They blame me for their hunger, but the passive way they gave into Marlene's bullying infuriates me, and I refuse to do it. I wait. We all wait. It is silent for five, ten, fifteen minutes. Then she comes over to me and screams. I don't remember what she says, but I start talking as soon as she stops. I tell her I cut my arm with a fork, my nails, a sharp wedge of my mother's broken wine glass, a scalpel from art class, the bricks that lined my father's tennis court. I tell her I stole from my parents. I tell her I slammed my mother's hand in a door because she threw her wine glass at me. I tell her I tormented poor defenseless girls at Berkshire, but by this point on the confession meter I might as well be admitting to smoking a cigarette. She questions me about men and sex. I tell her I am not a virgin. I tell her about Shane, about the bookish runner, about the dope head, about a drunken night when I followed a thirty-five-year-old Australian home from a bar because his accent reminded me of Crocodile Dundee. I was fourteen. She asks what I remember. All I remember are spurts of information, flimsy cardboard puzzle piece flashes: his hands tracing the outline of my barely developed body, burrowing into creases I didn't know existed and teaching me sensations I'd never felt, the dirty futon mattress, his whiskey breath, and coarse whiskers—something I hadn't felt against my cheek from any man other than my father, waking up naked, his tan body across my stomach, his tight white bikini briefs tangled mid-thigh. I tell her he dressed me and carried me to a cab. Marlene insists he raped me. I argue. She says I am in denial. I say that if he had raped me, I would feel different. She says not necessarily.

"Who else?" she asks, as if she knows I am keeping something from her.

I tell her a little about Elliot. She says—that's rape, sweetie. I cry, and she yells her argument as Karen Carpenter sings about finding a place where there is room to grow.

A door opens, and two unfamiliar counselors carry large sofa-like pillows into the room. They place a pillow and a pair of white socks in front of each person. Marlene says the next exercise is called the Pillow Pound. She says we must wear the socks on our hands to prevent chafing our arms and hands and to soak up the blood if we do open the skin. She demonstrates, thrashing her body up and then down, her mouth open, hollering. She says to imagine all our shame, all our sadness, and beat it away, destroy it. Hot moisture fills the room and the stretch of window panel on the wall is foggy. It's black outside, and all I see are trees and darkness, a dense layer of still forest. I look at Sam, and his face is white, his lower jaw hanging open as if he has just dozed off on a long car ride. I look at Jake, and he pretends he doesn't care. I know he does, though. Music blares from the speakers, this time Simon and Garfunkel.

The music is loud, and I see people begin to pound and scream. Words crash through the air: *fuck you, I needed you, don't touch me, I hate myself*. My hands are sweaty and cold, and my face is hot, and all I can do is listen to the wind roaring outside the barn. I think back to Connecticut and miss my father and wonder what he'd think if he saw me here, wonder if he knows where he sent me.

Marlene walks by and kneels in front of me screeching. All she can do is screech.

"Come on, Alison! Are you going to tell me all you are is a pretty face, unaffected, numb?" Her eyes burn through me, and she continues. "Are you proud of the way you behaved, whoring yourself all over New York City! Look where it got you! It got you fucking raped. You don't want to believe it, but it did! Your father didn't love you, so you went searching for a man that would, didn't you?"

"Fuck you. You don't know anything about me. I was not a whore, and I don't belong here," I say, but my voice is soft, barely discernible in the din of the room.

"You're not normal! You couldn't make it in the real world—remember? Life was too tough on Sutton Place!" she yells. Flecks of foam dot the corners of her mouth.

"Leave me alone," I whisper.

Marlene comes in close to my face and spits more insults at me, antagonizing me with information I didn't know she had. She calls me a whore again and again; an alcoholic, a drug addict, a thief, a spoiled, ungrateful brat. She speaks about my mother's alcoholism, my father's demeaning and incessant criticism. I wonder how she knows these things. Her cherry-red face is a foot from mine, and finally I break. I feel like Marlene has conjured the devil inside of me, and I begin to howl and shrill like an angry monster rising from a deep black lake. I pound the pillow as if my fists on the stupid seersucker fabric could really destroy my sadness. I am crying and yelling. I feel like a crazy person, a fucking lunatic.

And still music plays. It blares from the speakers. It shakes the room.

My screams collide with the singing words ricocheting off other screams and merging into one indistinguishable discordant noise that echoes through the room—I don't know what I'm saying, but I'm angry. I am deeply sad. I cry and yell and fall to my side in hysterics. Jake screams next to me. His face is streaked with mucus, and his eyes are a reddish black and surrounded by ruptured blood vessels, and he is groaning in between cries, and he is what I imagine a dying raccoon must look like. The worst part are his gloved arms, now bloody, forming two right angles that frame his head on the pillow. Everything seems awfully wrong to me. In the midst of my hysteria a thought floats through my mind: *I can't imagine this is legal.* Marlene comes to me and rubs my back. I want to go home.

Ms. Slywitsky enters the room. She is a bony woman over three heads taller than Marlene. She has flowing blond hair, but today it is wrapped tightly in a bun. She says it is time for the next exercise, an exercise of reflection, a walk of solitude. She motions for us to get up and follow her. A dirt path laid out years ago for the first Truth Workshop in the early seventies winds itself through dense woods. I see weeds swirling the trunks of old trees, tightly wrapped and beautifully green, then I stare up at the dark sky. It is black, much darker than the sky at home, and it stretches tautly above me like slick leather around a woman's thigh. As we walk I feel it closing in on us, slowly, like a tent collapsing, and it seems that there is no escaping this newfound sorrow.

"Are we ever going in? It's cold as hell out here, and it's so dark I keep tripping over these tree roots," Jake says. The cold air has calmed him, restored some of his cocky invulnerability, but the redness of his eyes and the way he holds his hands at his sides—curled, careful, as if they hurt—belies the brashness of his words.

"We don't use words such as 'hell' here, Jake, and we will be walking up to the dorms momentarily." Ms. Slywitsky glides across the uneven ground like a nun on her way to mass, calm and placid. We follow her in a silent line like acolytes, emptied and unquestioning.

We are not allowed to speak while we are in a Workshop, so we are silent in the dorms as we get ready for bed. I can hear someone locked in the bathroom stall whimpering. I think it is Jinny. The dryer is on and tossing clothes with metal zippers in circles. The other students know that a Workshop is taking place and they observe our silence respectfully, whispering to each other, or reading noiselessly in bed.

There will be more exercises tomorrow. We will meet on the Mill-house deck at 6 AM. If we are late, we will be locked out and sent to Dishes. We will not graduate to the next level. I set my alarm and close my eyes hoping for sleep.

Though for months afterward I was too ashamed to talk about what happened inside that barn, I did realize that the Truth Workshop was my first introduction to the reasons behind my depression. For years I had known it was there, felt it weighing me down, but had never fully comprehended where it grew from. Instead, I'd categorized my behavior as typical teenage rebellion. Now I had a vague idea that behind the sullen thoughts and anger was something more potent. The Truth Workshop was also my first major encounter with the brainwashing I would experience in various forms over the next two years. "Cascade Tools," like the Pendulum, would be handed out in workshops as precious tickets to the next level of personal growth, snippets of hypothesized information presented to us as hard facts. We as naïve teenagers assumed it was all documented and proven. We presumed that these adults who preached to us had degrees and experience—that the Pillow Pounding, the Running Anger, the Visualizations, the accusations, the verbal abuse and indictments were all the product of some program that had demonstrated results. We thought these adults knew what they were doing—that they knew how to live better than we did. Now I realize all they had were struggles with their own addictions.

14

LOWEST FORM OF LIFE

My parents visited me for the first time in July, five months after I arrived. My mother was stiff, her clothes flawlessly ironed as usual. She stood tall, wearing a white dress shirt, a cream cashmere cardigan, and black wool slacks. My father laughed nervously every few minutes and fidgeted with his wedding ring.

"Hot here," he said.

"Yes, very humid," my mother said, making a fan gesture with her hand.

We didn't know how to greet each other, so we stood in the dreary visiting room like guests at a cocktail party. Finally my mother pulled me toward her and wrapped her bangled arms around my torso. My body remained stiff while she squeezed tightly as if to revive me into loving her.

"I'm supposed to confess to you," I said against the soft cashmere of her shoulder.

"Confess?" my mother asked.

"Yes, that's what they call it here," I said, pulling back from her. "I have to admit all the things I've done wrong, so we can start with a clean slate."

"What is the point of that? The past is past," my father said.

"I don't know, Dad, but they tell us we have to do it, and if you don't follow the rules here, you do manual labor or scrub dishes," I said, bitterly.

"Go ahead, sweetheart," my mother said taking a seat on the visiting room couch.

My father took the large throne-like chair at the end of the room and placed his hands on its wooden arms, as if he were King Lear awaiting his daughter's speeches of adoration. Instead, he received my list of confessions: the sex, the drugs, the lies. I had to admit to stealing from them, to pawning my mother's jewelry and fur coat for drugs, and taking hundreds of dollars from my father's wallet. I had succeeded in lying coolly and easily without remorse, shame, or guilt for years, but sitting there faced with the admittance of such unconscionable acts to the man I had worshipped for so many years tore into me like a nail ripping the sole of a foot.

"Didn't we give you a nice life?" my mother asked. I'd been watching her face grow slowly paler over the course of my recitation. "All the opportunities in the world?"

"You make no sense, Alison, none! Most people would give their right leg to have your life, but you'd rather be a bum! That's all you are—a lazy bum, a druggie, part of the lowest form of life," my father said, furiously rising from the chair, the vein in his forehead throbbing.

My parents and me on one of their visits

My heart was crushed under the weight of his words. With a few simple sentences he'd managed to eradicate the minute amount of dignity I still held onto. My father had this kind of power over me.

"I'm sorry, Dad," I said. "I wasn't thinking straight. I made mistakes."

"Mistakes?" he yelled. "You're going to give me a heart attack. You're going to kill your own father, Alison. I can't do this. I want nothing to do with you." He turned to my mother. "I'll be in the car, Rosemary."

"He's upset, Alison," my mother said, watching him walk from the room. "He's very disappointed."

"Join the club," I said.

"What's that supposed to mean?"

"Nothing," I said. "I know he's disappointed, Mom, so am I. This isn't exactly where I'd hoped to spend my teenage years—in this *freak* school." I was crying.

"Dear, we will get through this. We want you to get better. It just takes time, I suppose."

I nodded, accepting her explanation as if it were that simple. As if one could just confess fifteen years of lies and deception and start with a clean fucking slate.

"Have you given any thought to the things Dr. Milton said?" she asked.

"I don't know. I just don't know anymore," I sniffled.

My mother handed me her handkerchief, and I wiped it across my face. It smelled like her, like the giant ostrich coat I once swam in as I begged her not to go out, like her scarves I borrowed during games of dress-up, her linen pillows, her hair. For a second I remembered what it felt like to love her.

"Ali, I am having trouble understanding why you did these things. Can you help me understand?" she said, leaning in as if I were going to whisper the answer in her ear.

I didn't know what to say. I thought of my room back home, about my friends. I missed my friends. I missed Aiden. I thought about the drugs and the cutting and the sex, and then I thought about the Truth Workshop and

Marlene's red face screaming at me and Jake's bloody hands, his bruised face, his chest heaving up and down, convulsing next to me.

"Because I was angry and alone and nobody understood me, and a person can't survive like that," I blurted.

"I don't understand. You were never alone. We were always there for you," she said.

Now I was hysterical, and I could tell I was terrifying her just as I had as a child, years ago. She dabbed tears from the corners of her eyes over and over again, staring vacantly into the green woods until I managed to reign in my hysterics.

"You'll never understand, Mother. It's 4 PM. You should go," I said.

She got up, kissed me gently on the forehead, and left. I stayed in the room, noticing the approaching fog headed our way. Dense fogs often appeared without warning in the mountains, hovering behind the peaks as we crept about, always knowing that at any moment they might close in on us, leaving us trapped indoors for days. Curling into a fetal position on the couch I began bawling into my hands. A moment later, I heard a knock at the door. It was a counselor named Katherine. She must have heard me crying from her office next door.

"Are you okay?" she asked, sweetly.

"Yes," I said, into the couch pillows.

She walked around the table and sat down next to me.

"Can I hold you?" she said. That's what they did at Cascade. The counselors held the students like babies and the students cried.

I didn't answer. Her arms came around me and sort of pulled me toward her. I fell into her lap. I didn't have the energy to fight. She smelled of wood smoke and lavender, and she held me tightly while stroking my hair, which made me cry even harder.

"Did you find out some bad news?" she asked.

"No," I said, shaking my head.

"Well, why all the tears?"

"I don't know. Everything is just so wrong right now," I said.

"Well, that's why you're here, so we can help you fix it all," she said, kissing the top of my head, "but you have to let us help you." And she held me until I couldn't cry anymore, and I fell asleep in her arms.

In September, the skies finally made good on their threats to open up and surge forth. It rained for nine consecutive days in vast thundering sheets, interrupted only by the briefest intervals of dryness that never lasted long enough to soak up the muddy puddles in front of my dorm. The trees and the grasses were permanently lit with that electrifying post-rainstorm green, and everything was dowered with extensive detail and color. The tips of grass blades poking through the dirt, the grain of the tree trunks, even the normally low stream rippled with force, glinting in silver half bubbles. I awoke most mornings to what sounded like millions of tiny hands tapping my window glass and wore a yellow hooded raincoat and perpetually soaked Converse sneakers, day after day.

Mira introduced herself to me on one of those rain-washed September days. When I had arrived at Cascade, I'd been given a Big Sister, an older student who was meant to befriend me and become a sort of role model or hero. My first Big Sister was a hyper, clammy-handed thirteen-year-old who pathologically embellished every story she told. We didn't get along, and when she graduated that summer I was assigned a new Big Sister, Mira Ito.

Mira was half Japanese, shorter than I was, with a high waist, long legs, and honey-colored skin. She had thick, shiny dark brown hair, a chubby yet pretty face, and a sort of perpetually bohemian air about her. We ended up at the same breakfast table one morning. It was a Sunday, and I was silently devouring blueberry pancakes when she began speaking to me. Eating was the only vice I had left at Cascade, and I depended on it, voraciously downing seconds and thirds at breakfast, lunch, and dinner. Sunday brunches were our treat of the week. Warm cinnamon buns swirled with vanilla icing, onion or sesame or poppy bagels with scallion cream cheese, waffles, chocolate chip pancakes, perfectly greasy hash browns—the list was endless, and I would return to the brunch counter four or five times in a sitting. By the end of my first year, I had gained twenty-five pounds.

"You're from New York City, right?" Mira said.

"Yeah," I said.

"Me too," she said.

"Really." I forked a bit of pancake into my mouth.

"Well, I was living there before I was sent here. I'm not exactly from there," she clarified, shrugging her shoulders.

"Where else have you lived?" I asked.

"I've lived in Canada, Japan, and New Mexico," she said, casually.

"How did you end up living in all those random places?" I asked, as the other two girls at the table got up.

"My mother basically followed shoddy men around the world."

"Oh."

"My stepdad lives in New York; actually he's not even my stepdad, but he feels like it. He's just my mom's boyfriend. They're not married yet, but soon, I hope. He's a lot older. Seventy. I don't really get it."

"Maybe she loves him," I suggested.

"I think she loves him okay," she said. "Anyway, I have to get going. I'm supervising Dishes tonight. I'll see you later?" And she was off, carrying her tray and a small pile of trash she'd collected from the tabletop.

I was drawn to Mira right away. There was something very comforting about her. On one hand, she was damaged and full of fear, yet she still seemed to possess a solid strength of character I wasn't familiar with. I latched onto this immediately, and almost at once began exposing myself to her like I hadn't to anyone in years. Her charm and warmth; the softness of her voice; her pudgy, persistent fingertips that played with my hair—all seemed to hypnotize me into a state of unquestioning ingenuousness. Within the safety of our dialogue, feelings and memories I would have otherwise never discussed suddenly rushed forth. People said we were consumed by each other, that when we were together, nobody else existed, and it was true.

Before I met her, I spent my free time at the school in stupefying boredom, languidly moping around the House where we spent most of our free time, playing monotonous and definitively lackluster games of gin rummy, or pretending I was interested in learning how to crochet and drifting in and out of consciousness while Jake tried to teach me how to sew a yellow bag. Mira made Cascade tolerable for me. She had been there a year and a half when we met and was graduating in seven months.

We found new ways to entertain ourselves. We roamed the campus with cameras and black-and-white film, doing our best to take interesting shots with the limited access we had. We photographed each other and gave students candy bars if they'd agree to pose for us. We wrote poetry and read our poems passionately aloud in the back of the computer lab. Sitting cross-legged, clutching Styrofoam coffee cups filled with juice, I'd imagine we were at a chic poetry reading in an East Village coffee house, reading our work among notable writers. Mira wrote mostly about her mother's abusive boyfriends and junkie friend Hadyn Michaels, who'd overdosed on heroin a year before she was sent away. The poems were vague and eerie, laced with that frustrating yet equally intriguing quality that engages a reader, yet leaves them wondering what the hell the poem is really about. I was infatuated with Mira's writing and her life in general. I couldn't get enough of her stories about the Buddhist monastery in Japan, the drug ring operated within it, the kidnapping of her little brother, or her days as the youngest hippie in Haight-Ashbury.

While at Cascade, I had become obsessed with the power of atrocity, convinced that great writing depended upon great hardship. Aware of how wrong it was, I couldn't help but envy Mira for the never-ending cornucopia of tragedy she had in reserve to pour onto the page. I longed to soak in a deluge of calamity, to gain power and perspective from unbearable agony. I was convinced my monotonous, whiny poems, limited in theme to my mother's drinking, my teenage rage, and my rather muddled critique of upper-class New York, were insufficient. So I switched to fiction, and soon I was writing stories capable of driving even the most cynical reader into a four-day depression. This I felt was true success.

During my two years at Cascade I had faux love affairs with various boys, but by far the most unrequited adoration I ever felt was for a boy named Jamie Steckleson. I worshipped him from the day he moved in with his oversized guitar covered in rainbow stickers. Some were bands he had a fondness for—Grateful Dead, Nirvana, Talking Heads, the Allman Brothers, Pink Floyd, Phish. Others were bumper stickers saying *Be Green* or *War, what is it good for?* The counselors made him peel all the stickers off, putting masking tape on the ones that refused to wash away even with Comet and a Brillo pad.

Jamie's voice had a unique and reassuring huskiness to it that I felt deserved a place in musical history with Dylan and Young. Not only could he sing, but when he played his guitar, you could feel the soul of the music in your gut, and often the intensity of his performance brought me to tears. He'd sit in the center of a circle of girls and boys in lotus position, a tie-dyed guitar strap hanging across his broad shoulder, humming and plucking away at the strings, until he was centered enough to break into song—*My, my, hey, hey. Rock and roll is here to stay. It's better to burn out than to fade away.*

I'd lie flat on the gritty cement, eyes closed, arms stretched in a diamond around my head, lifting my sweater to reveal a sliver of bare stomach I hoped Jamie might notice, though his eyes were usually closed when he played. Some kids would dance, skip in circles, or simply stand upright and sway back and forth like tropical palm trees. Even Lara Brown, who claimed to detest classic rock, would sit in one of Jamie's circles and sway gently back and forth with that glazed, lovelorn look across her face. Once in a while I'd even catch her singing along.

Mira didn't like Jamie, nor did she approve of me hanging around him. She said he was an inadvertent womanizer. She thought herself too good for "his type," and she was right. Since we weren't allowed to talk about boys directly, we found ways around the rules. "Cutting Corners" was what the Cascade counselors called this, but it was more tolerated than breaking the rules directly. We'd lie in bed at night and set up "hypothetical" situations without identifying the characters as more than boy J or girl A.

When Jamie passed me in the cubby room, he'd cock a shy smile from the corner of his lips and surreptitiously brush his hand against my hip, rubbing a finger slowly up and down my navy corduroys. His fingers were big and strong, good for manipulating guitar strings and tightening cords. He flirted with all the girls and probably touched them all in the cubby room, but of course, I liked to think I was special. Beyond the physical attraction, I cared very much for Jamie. He was one of those people you find yourself profoundly attached to within seconds of meeting. I could tell he knew more about the world than I did and this turned me on. He knew people, knew their motives, their tricks, and their stupidities. Oddly enough, I don't remember all of Jamie Steckleson's story, but I know he had a sordid past: a dead mother and an alcoholic father who had some ailment severe enough to necessitate hospitalization a few times a year.

Jamie was especially desperate to stay connected to the outside world, afraid of getting too comfortable at Cascade. With the little access we had, we did our best by scanning the newspapers for the weather outside our rain-soaked campus, rummaging through acceptable magazines in search of concerts or musical gigs, new albums, or star gossip. During the winter, Jamie and I went sledding on garbage bags and designed intricate sculptures out of snow, naming them like installations at the Museum of Modern Art and mourning for them when a warm day came and melted them away.

As with most good things at Cascade, our friendship soon came to an end. We were placed on Bans for going over our Time Limit of twenty minutes, and then four days later, Jamie was caught breaking into the nurse's office in the middle of the night, probably in search of pills. He was escorted to a lock-up the following afternoon, and we didn't see him again. For years after I left Cascade, I listened to the radio with a certainty that one of these days I'd hear that pained and husky voice croon from the speakers. But I never did.

After Jamie left, I found other boys to obsess over. These contrived love affairs gave me something to do as I scrubbed pots and pans for hours or lay in bed at night, unable to fall asleep.

15.

NOTHING TO
TALK ABOUT

By October my Family had moved up in the program. We graduated Beginning School and entered Middle School, where we were introduced to a new counselor, Jerry Thompson. He was a tall, lanky man with blushing cheeks, thick, mossy eyebrows, and a balding head. His laugh was genuine, his smile warm and welcoming, his hands soft and large like the paws of a bear cub, and his presence capable of instilling immediate comfort in even the most frightened student. He seemed to care for us immensely, less out of duty and more out of a sincere desire to help. Cinnamon cologne was always splattered haphazardly across his neck, soaking the collar of his signature blue or green polo shirts, and it was impossible to hug him without getting a mouthful of the pungent smell.

He suggested our first meeting take place by the duck pond. I agreed, relieved that I would no longer be forced to sit in the dreary wood-paneled office for two hours a week. It was fall; leaves were gathered in bunches along the pond. I kicked into them as we walked and they fell slowly, rhythmically, wavering back and forth like a seesaw. Their slowness comforted me. Jerry didn't make me speak like the other counselors had, didn't make me feed him trained answers I had been stuffing down the throats of doctors since I was five. We walked in a relieving silence and I allowed myself to be hypnotized by the quietude of the afternoon.

"What do you think of Cascade?" he said after some time.

"I'd rather be home," I said.

"But we know that's not an option. So assuming you have to stay here, why waste your time?" he asked.

"What do you mean?"

"I mean why not work on yourself. Make yourself happier, a better person. It says in your records that you've never worked in a forum, never Run Anger."

"I have nothing to work on," I said.

"Alright," he said, and pried no further.

On Saturday mornings the student body was divided into Work Crews. Sam, Jake, Lara, Marilyn Freedman, and I were assigned the copse of oak trees surrounding the track where the obsessive-compulsives, weight fanatics, and wannabe athletes treaded hour after hour. Their feet lacking proper support, shins splitting, and heels bloody, they'd plod on until physical exhaustion overtook them. When I gazed out the cafeteria window during most meals, I'd see the same stick-figured girls running around and around in their flimsy gym shorts like they'd just run off the page of a child's drawing. These were the girls who circled the salad bar for twenty minutes at the beginning of every meal picking raisins and crunchy Chinese noodles from the mixed salad like mother birds might collect bugs to nourish their young.

Marilyn Freedman was one of Jerry's students as well. She dressed in flowing white pants, Birkenstocks, and wool socks, and prefaced everything she said with *The fact of the matter is.* Her hair was cut short, thin and dark brown, and she could sing "Me and Bobby McGee" nearly on par with Janis Joplin. I'd taken an acting class with her in New York City a few years before, and we'd recognized each other immediately upon my arrival, but were put on Bans for my first two months because the school feared we might tell War Stories about the old days. What they didn't know is that we had been enemies in the old days. I think we'd both seen the same ugliness in the world we came from, but dealt with it differently. She went into the depressive low and I into the manic high, and we hated each other because we each recognized the other's act. She'd come to Cascade after two weeks of living on the streets in Texas and a brief visit to Silver Hill. Many people thought Marilyn was insane; they didn't get her. They couldn't see that her boisterous antics and depressive lows were simply a mechanism of protection from the very real self-loathing inside her.

Strangely enough, I recall my Saturday Work Crews with great fondness. In my memory the sun was always shining, and the fall leaves always crunchy and dry and smelling of earth, sap, and pine, and there seemed an overwhelming yet momentary presence of contentment. There was a camaraderie among us on those afternoons, as we raked and piled and bagged, a union of endurance and shame, both for what we'd been through and for what we knew was yet to come.

During my stay at Cascade, there were moments when it felt like a utopian universe, moments like those when I'd forget what it was or why I was there. One afternoon as we lay in the pile of leaves soaking up the late morning sun, Sam turned to me and said, "Here I am locked in this therapeutic rehab, this shitty Nazi-like school, and I can't help but feel this is the happiest I've ever been."

I looked at him.

"But this isn't real," I said. "Life doesn't work like this, with bedtimes and sugar limits and everyone medicated and crying whenever they want."

"Why not?" he said, widening his eyes.

What I realized later was that Sam didn't want to live in reality. He wanted to stay young and adorable forever. When he left Cascade, he tried to pick up his skateboarding career again, but was hit with the troubling verity of age. He lacked the agility, the quickness of youth, and no label wanted him. I am not sure what he is doing now, but last time I saw him he was selling hot dogs in the parking lot of a department store.

During my second meeting with Jerry, we walked around the pond chatting about frivolous things. We passed a rotting barn that housed five or six old canoes. Rusted hinges held the door upright as it flapped in the wind like an abandoned farm gate. We tried to peek inside, but the windows were webbed with spider's silk, and milky dust layered the ledges. We stood there for a moment.

"Guess these canoes haven't been used in a while," I said.

"Nah, these old guys haven't been dragged out in years," he agreed. "The newer ones are with the kayaks in the wilderness shed."

"I always wanted to learn how to do that kayak flip," I said.

"I'll teach you this summer, kiddo," Jerry said. "But listen, we have to cut the shit and start looking at you. Kayak flipping is not what your parents are paying us to teach you."

"What are they paying you to teach me?" I asked.

"Well, they're paying us to teach you how to lead a successful life, I suppose."

"Well, teach me then," I said, pissed off.

"Alison, don't be a smart aleck. You have to start working on yourself."

"Jerry, I really don't have anything to talk about," I said. "I told you. I don't think I'm sick. I don't think I have a disorder. So I get a little depressed once in a while, about what, I don't even know. My parents didn't molest me. I've always been fed well, and I've never been locked in a closet. I don't know what you want me to talk about."

"Well, we could start here: alcoholic, unstable mother, cutting, slamming head into walls, heavy drug use, robbing parents. These are just a few things that Tina had noted in your file," he said, reading from a coffee-stained notebook. "You wanted something to talk about. Looks like I found it."

"Not really, because the thing is that I don't care. It doesn't affect me. I mean, it would be nice if it had all worked out differently, but it doesn't really faze me. I don't even think about it," I said.

"We'll chat about it in Forums tomorrow," he said, walking away with a wave.

The following afternoon was the first time I spoke about my Issues in a Forum. Twelve students, including Mira, stared in my direction as Jerry questioned me. I did as I was told, partially because I trusted him and partially because I was starting to believe in Cascade. I closed my eyes and placed my head in my hands and pictured my mother in her white lace nightgown wandering through the living room as she often did late at night. Her long feet were bare, and with each step I could hear the moist sole of her foot parting with the wood floor and reconverging with it a moment later. Sometimes she'd stop for a few minutes, cross her arms, and seemingly deliberate about something very important as she stared vacuously through an object. Other times, her cloudy eyes would glaze slowly over the entirety of the room as she turned in circles again and again like plastic teacups at a carnival.

She never knew I was there, standing under the frame of the French doors that connected the dining room to the living room. She never knew I was spying on her: a little girl in bare feet and a smocked cotton nightgown, a ten-year-old in pajamas, a teenager in boxers and a T-shirt.

I said all this or something like it out loud in the Forum, and the next thing I knew I was Running Anger, screaming at my mother with all the rage in my body. I tried to remember what it felt like when she threw a wine glass at me or grabbed me with her sharp manicured nails. I tried to remember what it felt like when she looked at me as if I were a monster. Jerry talked me through it. He prodded deeply into me, peeling back layers one by one, like flaps of skin being pulled from a lab animal. I must have yelled for fifteen or twenty minutes. When I finished my throat was raw, and I could taste metal in my mouth. My nose had begun to bleed. For a minute, I sat there feeling—something, I don't know. Emptied. Hollowed out. But soon this emptiness began to fill with guilt, guilt for publicizing these things about my own mother. It was the same guilt I would feel in my late twenties when I began attending Alanon meetings. I never spoke, but just sitting in that room was acknowledgment enough. My mother had always told me, *Ali, whatever you do, never air your dirty laundry in public,* and there I was airing not only my own dirty laundry but hers as well. Twisted as it is, I felt as though I had betrayed her.

"I'm so proud of you, Alison," Mira said as she rubbed my back.

"Thanks," I managed to say between sobs.

"It wasn't that hard, right?" she said with gentle sarcasm.

"Yeah, right," I said, and we laughed weakly together.

16

GOOD INTENTIONS

By early December, my second parental visit was two weeks away. This was my first off-campus overnight visit, and I walked into Jerry's office for our usual session, eager to discuss the privileges I'd be allotted. The weather had grown too cold for us to continue our walks, and we were forced inside.

"Can I listen to music?" I asked.

"I'll make you a deal. You can listen to unacceptable music and watch one hour of television, if you confront your mother on her alcoholism. It really needs to happen," he said.

"I can't do that. Plus, she'll deny it, so what's the point?" I countered.

"The point is so that some day you can have an open and honest relationship. You're angry at her. She hurt you, and you have to tell her this, or you'll never be able to forgive her."

"I'll try, but I can't promise anything," I said.

"C'mon, kiddo, be brave." He put his hand on my shoulder and squeezed.

A knock came on the glass door, Jerry stuck his thumb to his nose and wiggled his fingers. I knew it was Sam. He stood on the other side of the door, head cocked to one side, tongue protruding limply from his mouth. Sam always looked as though he should be in diapers dragging a worn blanket behind him like some cartoon baby. Jerry beckoned him in.

"Hey, Sammy. What up?"

"Same old. Sad, I guess," he said, Thorazine-shuffling across the carpet.

"Why? Anything unusual happen?" Jerry asked.

"Nah, life, you know."

I could see Sam was holding back tears as he spoke. There was a tautness in the skin around his eyes, a tension in his clenched fists and the frantic batting of his lashes. His cheeks were bright red, and he rocked from one foot to the other while fiddling with the loose strap of his backpack.

"Well, we're pretty much finished," I said. "You can take the rest of the appointment."

"You sure?" Sam said, avoiding eye contact.

"Yeah, yeah, go ahead," I said.

"We'll talk again before your visit, Ms. Weaver, but think about what I said," Jerry insisted as I stood up to go, tapping a finger to the side of his head.

A few minutes later I walked by the glass door and saw Sam cradled like a newborn baby in Jerry's arms. Sam refused to cry with anyone but Jerry, not even in Forums and rarely in Workshops. His navy-and-white, unlaced sneakers rested near Jerry's thin leg, one on top of the other, the frayed and muddy jean cuffs bunched at the ankles, rippling unevenly over the sneakers. Gray sweatshirt sleeves were stretched over his hands covering his wet face.

My parents arrived to pick me up for my visit precisely as the second hand on Mira's watch hit eleven. I hated them for it. Walking along the wooden bridge, my mother hanging on my father's elbow for support across the icy path, they looked like perfect parents. Seeing me through the glass window they smiled, waving. I knew that under my mother's dark Dior sunglasses, her eyes were welling up with tears. She always cried when she saw me after a long absence, even when I returned from a weeklong sleepaway camp at eight years old. It was as if she feared I would discover the life I wanted somewhere else, and she would lose me forever.

"Hello, dear. Is this your friend?"

She bent forward for a hug, and I remained stiff. Her head rested on my shoulder for a moment, and I felt the slow swallow of tears. She seemed miserable. And what made it so much worse was the effort she put forth to appear happy.

"Yes, this is Mira," I said.

"Nice to meet you, Maria," my mother said, shaking her hand.

"Mira," said Mira.

"Oh, right," my mother said. "That's an unusual name."

"What happened to your legs?" my father pointed toward my legs as if I didn't know where to locate them.

"Yeah, I've gained some weight here," I said.

"It's nice to meet you both. I better get going. I have class," Mira squeezed my hand as if to say—be strong.

My father placed his arm around me and clutched my shoulder. He smelled wooly, like he always did in the winter. I guess he must have decided to give me a second chance. We never spoke about it. One night he

Dad and me on a visit

answered the phone when I called home and began asking me questions about my classes.

Later that day we would take a picture in front of a gaudy reproduction of Van Gogh's olive trees in the lobby of our hotel. It is that picture that is included on my Cascade yearbook page. My father's mouth is in mid-smile, as though he is saying something to me or to my mother who is taking the picture, something off-color and sarcastic. He looks genuinely happy standing there with me, and I too look happy. We both wear beige corduroy pants, and he wears a wide collared tattersall shirt with fading blue and brown checks and worn elbows and cuffs. A green wool sweater is tied around his neck.

This was one of the last pictures ever taken of us together.

As my parents and I walked across the bridge toward the car, my mother grabbed my arm, afraid she might trip. I hated how weak she was. From behind we might have looked like a happy family, arms linked, rejoicing at our reunion. I imagined the other students staring through the house windows and watching my parents and I revel in our wholesome family bliss.

Dinner was long and tedious. I was amiable, appeasing them with frivolous small talk for the first half of the meal. But the act of pretending eventually became more taxing than accepting the bleakness of the situation. "You know, we're supposed to be talking about our family issues," I suddenly said, having mustered up enough nerve to confront my mother.

At the time my decision to tackle this issue came more out of my desire to please Jerry than any serious longing to build a relationship with her. Second chances or no, I knew my father would never be proud of me again, and Jerry was the first of many men whose praise I would use as a fix to mollify the pain surrounding his disenchantment.

"We know, but we thought we'd let you get comfortable first. We're ready when you are," my mother said.

"I have some things to say, and they might upset you," I warned.

"Okay," she said.

I proceeded to confront her on her alcoholism for the first time in my life, and she denied it until her throat was dry. She said that I was being *silly*, and that just because she may get a little *tight* once in awhile did not mean she was an alcoholic. Any and all evidence I supplied was deemed untrue because it never happened. My father had no comment on the subject other than *she seems fine to me.*

I thought then that she was lying as she often did, denying the truth of her alcoholism like alcoholics do. But I realize now that her perception of herself was drastically skewed. The cycle of pain that fueled her disease was so compounded and layered that even now I find it difficult to pick apart or label it in any concrete way. Looking at her was like looking through a kaleidoscope, millions of colors and shapes spinning, form and pattern constantly changing. She must have known she had a problem or she wouldn't have lied. She wouldn't have fibbed about vodka being in her Bloody Mary, or hidden wine glasses behind picture frames when I entered the room or purchased a sparkling juice called Aqua Libra that was identical in color and consistency to white wine, so she could claim to be drinking that at eleven in the morning. But at the same time, because she had friends whose behavior leaned more toward the stereotypical definition of a drunk, perhaps she couldn't recognize the severity of her own problem.

I got up to use the bathroom after the confrontation, and when I returned, my father was gone, and streaks of tears glistened in my mother's foundation. I hadn't reached for her hand in years, but as I slid into the booth I grabbed it and held it tightly for a minute.

"Ali," she said. "If my drinking upset you, I am sorry. You were young, and maybe it seemed like I drank a lot more than I really did. Children are sensitive to these things."

"Yeah, maybe, Mom," I said.

"I am sorry," she said again.

"Why were you always so upset when I was young?" I asked.

"What do you mean?"

"You know. You cried a lot in the bathroom, late at night."

"You must have imagined these things, sweetie, or dreamt them. I don't know what you're talking about."

Her thin hands wiped at tears that now fell from her eyes in a constant stream, and it occurred to me that maybe I'd gone too far. Being the witness to my mother's heartbreak was both agonizing and revolting, and as we sat there in the Red Lion Inn, a mother and daughter staring across a table of half-eaten steaks, diluted sodas, and stained napkins, all I saw and all I would ever see was the heavy weight of fifty-five unfulfilled years and the deadening certainty that she would never escape. The life of a society wife had grossly disappointed her, yet she would never have considered leaving it. She wanted to be Mrs. William Merritt Weaver Jr. forever, no matter the price.

The following morning I found this letter slipped under my door:

Dear Ali,

I am writing this letter with mixed emotions. I am sad that you are so sad and that I caused you that sadness. That is the last thing in the world I would ever have wanted to do. But I am also happy in a way and relieved that we can finally start talking honestly. There is so much I could have told you but I never thought it was appropriate. I didn't want to frighten you by blurting out—I was sad through my entire childhood. I wanted you to have the perfect childhood—a perfect little setting—ha—I didn't quite succeed—did I? Well, more to come, to be continued as they say—

I love you very much.

Your mother

She'd clearly had good intentions by writing this letter. She was trying to open a path of honest communication. She even began seeing a psychiatrist back home. What happened, I don't know. I never received another letter like that one and we never discussed any of it again. Maybe the pain was just too much for her. Maybe she decided it was too late to change.

Later that evening, after we settled into our separate hotel rooms, my mother probably sipping wine in the bed next to my sleeping father, and

I enjoying the freedoms cable television offered, a girl came knocking on my door. She was a fellow student of mine at Cascade, also on a visit with her parents.

"I thought you might want to join me for a smoke and a dip in the hot tub?" she said. It was strictly against the rules to interact with any other Cascade students, smoke, or leave the hotel room alone, but at that moment nothing sounded more enticing than a cigarette. She flashed a pack of Camel Lights in my direction, and I grabbed my bathing suit and followed.

The noxious hot tub chemicals burned my eyes as we sat immersed in a mushroom cloud of smoke and steam. I hadn't used lotion in months because I hadn't been rewarded my Lotion/Perfume/Hair Product Privilege yet, and the scalding water hurt my chapped skin. At Cascade, everything was an earned right. From sharing toothpaste to riding a bike to eating chocolate—all must be earned.

"So, what do you think of Cascade?" the girl asked.

"It's bizarre," I said, reveling in the peaceful silence of the evening.

"I think it's a cult," she said.

"Why?" I asked.

"It has all the colors of one. Think about it. We're all taught to believe a certain set of values, live by certain rules, worship certain people. It's Waco all over again," she said, nodding.

"It's not a cult. It's just a weird rehab with excessive rules and regulations."

"Well, you say tomayto and I say tomato," she shrugged.

"Where did you get these smokes?"

"The gas station next door. I showed the guy my breasts in exchange for a free pack." She folded her head forward as if bowing to applause.

"Good thinking. I wish we could get a drink somewhere," I said.

"They don't sell alcohol at the gas station. There is a bar across the street, but I don't think we look old enough, and my breasts are good but not that good. They'll do for cigs but not liquor," she said, staring critically at her overdeveloped body.

"Yeah, what kind of hotel is this that doesn't have a mini bar? We could have gotten drunk off the liquor in there."

"A shithole in a shit city," she laughed, smoke streaming from her mouth.

The next morning my parents drove me back to Cascade. I remember standing with them in the parking lot, their silhouettes sturdy against the dim and foggy morning. Students passing us on their way to breakfast, linking arms and laughing, boys biking past or chasing each other.

"Bye, Al, good seeing you. Study hard, and it will pay off," my father said, lifting his hand for a high-five.

"Okay, Dad," I said, even though we both knew that A's from Cascade would get me nowhere as far as good colleges went.

"Goodbye, dear, we'll miss you," my mother said.

"If you'll miss me so much, then why don't you take me home?" I asked. One last try.

"You were out of control at home, Alison. You needed help. You remember?" she said.

There was little I could say to that.

As the sound of their car faded down Whitmore Lane, Enya played on the stereo in the House, solely for the purpose of making us sad. I could hear the faint cries of students sobbing in each other's arms, and I wondered what on earth one could possibly be crying about at eight in the morning. I wanted nothing to do with those cries and the therapeutic garble I knew was being ping-ponged back and forth. But the problem was that I didn't want my parents and their life either.

This particular visit shifted my perspective. Previously I'd been convinced that since my family history lacked any kind of serious trauma, an expression of sadness on my part was unjustified. After all, I was privy to a mother and father, each with four limbs, who could form sentences and were well-meaning, who'd sent me to school and bought me toys and took me on vacations. To complain, when I clearly had it better than most students at Cascade, struck me as selfish. I hadn't yet fully acknowledged the possibility that though my life was largely deficient of harrowing events, I had still been hurt.

A perfect childhood. A perfect little setting. I thought of my mother's letter. Had that really been what she'd thought she'd given me? Had it never occurred to her that I was unhappy? A warm wind gusted through the towering trees making me conscious of the desolation and remoteness of the place surrounding me. I was miles away from anything familiar, yet aware that not even the familiar would be of comfort now. My parents would be safe inside their Sutton Place apartment in a few hours. They'd be served dinner on silver trays in front of the evening news. They'd go to sleep with water pitchers by their beds and awake to freshly squeezed orange juice. They'd attend a party the following night and when asked about me they'd say, *A boarding school in Switzerland,* and laugh with relief, realizing how easy it was to draw silence like curtains over the shame.

The dense, humid air of tears filled the House as I entered it. There were always at least three or four people crying in various corners of the main room, sad for one reason or another. Mira caught a glimpse of me and turned in my direction. "How was it?" she asked, pulling me toward the sofa.

With all the words in the world, none seemed suitable for what I felt at that moment. She pulled me into her arms, and I could hear her heart beating and smell her sweet perfumed skin and feel her chest rise and fall as she stroked my hair. I leaned against her like a corpse, and closed my eyes. Then I felt another hand on my back; a big, pawlike hand. I opened my eyes to Jerry, smelling sweetly of cinnamon.

"You okay, kiddo?" he asked.

"Just sad," I said, and he nodded, kissed my forehead.

"Breakfast will be over soon," he said. "Why don't you go eat?"

I wasn't hungry but I got up anyway.

17.

BETRAYAL

The day I found out about Jerry, the temperature outside had dropped below freezing. Ice spiderwebbed across the corners of the dorm windows like an old lady's cracked skin, and we were awakened by intermittent crashing from the woods, as tree branches buckled under the weight of frozen snow. That morning, I waited for Jake by the nurse's station inside the bleak, wood-paneled House. I watched through the scratched glass doors as he slouched toward me in his bright blue down parka, baseball hat, and hiking boots. He kicked the swinging door open, shivering, and lifted a hand in a lethargic wave. Prior to class, everyone lined up outside the nurse's station for their morning treatment, which consisted of some sort of antidepressant, antianxiety, or antipsychotic pill. I refused medication because it frightened me. I was afraid it would turn me into someone else. And since I hadn't been firmly diagnosed with anything, they accepted my refusal.

Often, though, I stood in line with friends as they waited for their colorful combination. Jake and I stood silently. His face was red and crusty with peeling skin. His nose dripped a clear bubble, and he didn't bother wiping it with his glove or grabbing a tissue, just let it hang there waiting.

Lara entered the House from the door at the other end of the building, clad in a white puffy jacket and a gray wool hat. When we'd first met, she'd shown me some pictures of herself back home in Los Angeles. She wasn't supposed to have them but she'd snuck them past the search somehow, determined to hold onto a piece of the past. In the picture she wore dark black eyeliner, fake lashes, ripe red lipstick, and thick gold hoops. And I remember her telling me that she was always struggling to be seen, to be noticed—that was why all the makeup and all the anger. She said she felt invisible without it, and maybe she was right to feel that way, because now as I watched her move toward us she seemed almost ethereal. Her skin was washed out, her eyes so swollen from crying that they barely resembled the shape of eyes any longer, and her other features blended straight into her face without distinction. Cascade did that; it took so much from us that we all fused into one another like cattle. I lifted my right arm. She clasped onto my body from the side like a pair of tweezers. Her head rested on that flat space below the shoulder and above the breast, and I could feel her body breathe, in, out, in, out, slowly. We moved up in line, and Jake stepped forward for his turn, hand extended like a trained animal, palm open.

"Good morning, Mr. Forester. Are we feeling better today?" The nurse cocked her head slightly.

"Yes, Pat, thanks for your help. You always know exactly what I need," Jake smiled, flirting, always flirting.

I looked down at Lara and kissed her forehead. I didn't know what particular issue was bothering her that morning. I assumed it was the usual: her abusive father, the molestation, her sick sister, or issues with adoption. But I was wrong; today it was something else. This was something Sam had told her. Something Sam had been wondering about. He said he'd been feeling confused about an incident that occurred with Jerry. He said they'd been hanging out at Jerry's house one afternoon, lying around on the sofa, wrestling or talking as they always did. But he said on this day Jerry had taken his hand from an appropriate spot on Sam's thigh and slid it slowly into his crotch. He said Jerry let it sit there for a moment, almost cupping his *thing*. Sam had sat up and backed off, and in a nervous reaction he'd lifted his shirt and said *Hey, Jerry, look what I can do* as he began rolling his stomach, at which point Jerry leaned over and slid his hand down Sam's pants under the elastic of his boxers. Sam said he didn't know what to do, so he just sat there and kept talking, trying to skirt the obviousness of the situation as Jerry fiddled around inside his underwear.

He'd confessed all this to Lara, genuinely unsure what to do. She'd insisted they come forward. Sam refused, and Lara was forced to betray him. Most counselors didn't believe it, couldn't believe it because they were Jerry's friends. When Marilyn found out the following day, she expressed concern to Marlene, who responded with *You know it's hard Marilyn because sometimes people get confused about reality and they misinterpret certain things. Some people don't understand what is right in front of them. You have to be very clear on who is speaking and how seriously you should take them before you get too worked up.*

That night I sat in study hall replaying the story over and over again in my head—and though I am reluctant to admit it, trying to figure out a way in which the interaction had been misinterpreted. My assigned table for study hall that semester was between the glass door of Jerry's office and a row of thin windows. Boxes sat on his desk, some taped and labeled, others still open and half-full. Pictures of him picnicking with his students on his vintage yellow fire engine were still pinned to a corkboard above the desk that was lined with antique car collectibles. He walked back and forth removing pens and notepads from the drawers, pulling numerous psychology books off the shelves: Freud, Jung, Winnecot, and stacking them in boxes. The expression on his face was unmistakable shame.

We never saw Jerry again after that night. I suppose he was asked to leave immediately; it was all done very quietly. Few people really knew why he'd left. Some inquired, and when they were met with responses that were cryptic and vague, they ceased questioning and speculated in hushed whispers.

Sam blamed himself for the entire situation. He thought it was his fault that their friendship had turned from something so innocent into something so corrupt and dirty. He missed Jerry almost intolerably, at times wishing he had never opened his mouth, almost willing to endure the violation in exchange for the love Jerry had given him.

By fall, things had quieted down. The memory faded, his house repainted and given to a new counselor. Then more bad news came. A Cascade graduate killed himself, jumping forty stories from a high rise somewhere in Chicago. He had been one of Jerry's students. No connection was ever drawn or even speculated out loud, but the unspoken assumption in many of our minds was so pungent and crushing that you could feel its weight hanging in the room when the news was revealed.

Sam shut himself off after Jerry left, completely and entirely off. For months, he refused to speak in Forums, refused to cry or talk about anything at all. He began breaking into the nurse's office, stealing Xanax, crushing it late at night in the bathroom stall and snorting it. I tried talking to him, but he no longer responded to me, just shrugged his shoulders and pulled his soft, sweet lip into his flat nose. He seemed to float about the campus like a lost spirit looking for its grave, fingers hooked on belt loops, eyes cloudy and blank.

Jerry's cabin at Cascade, which was later razed.

18.

THE PROMISE OF A BETTER ME

Though Cascade bore only a tenuous relationship to reality, the defenses I had previously used to protect myself from the looming presence of life outside my head were the same that I utilized to protect myself from the grabbing hands of the school's ideology. And it was precisely those protections that failed me.

There are a number of reasons the school succeeded in breaking me. First, they brainwashed us. They used tactics employed in Hitler's Germany, Stalin's Soviet Union, and Mao's China: confession, total devotion to the group, regression to a childlike state, milieu control, need for purity, removal of privacy, guilt, isolation from reality, promise of salvation. At first I gave in out of desperation, out of exhaustion—because I couldn't spend another night scrubbing urinals and pots and pans and filthy rubber kitchen mats. Because my hands were blistered and my back throbbed. Because I couldn't take the incessant screaming or belligerent verbal attacks any longer. I just didn't have it in me. And because, ultimately, I became convinced that the school could save me. I wanted to believe in the promise of a better life, and the more convinced I became, the more I bought into their program and the harder I worked. I followed the rules meticulously, tattled on others I caught breaking them, and spent every Forum scrutinizing my own behavior. It was so subtle, so creeping, that I almost didn't know it was happening. But it was: I was becoming someone else entirely.

I will also say that—brainwashing aside—there was something else happening inside me. To be touched like the people at Cascade touched me roused some deep need to be loved and vulnerable that I hadn't previously been conscious of. It felt good to be treated tenderly and to have people hold me and love me unconditionally. And it felt good to grieve.

Once I returned from my overnight visit in Redding, an unpleasant surge of guilt enveloped me as I went through my daily activities. I struggled to suppress it, but the guilt continued to grow until it was so fierce and large in scope that I felt like I had removed a life from the earth. Logically, I knew that smoking cigarettes in the hot tub was not a sin, but my logic was in the process of a slow dissipation that would continue over the next year. At last, I Copped Out (a Cascade term for admitting to breaking rules) and was put on Level Three Dishes seven nights a week for the next three months.

I wore soiled shorts and loose tank tops but still turned beet red from the heat of the gargantuan metal dishwasher. I lifted heavy trays of chapstick-caked glasses into and out of the machine. I dug ditches and filled them and dug them again. I did all this with the conviction that it would lead to happiness. After all, wasn't it Siddhartha who said that life was suffering and the only way to happiness and truth was through this suffering? Sometimes I imagined the little black pearls adorning my arms and the dirty sweat that beaded my forehead and chest to be evil expelling itself from my body. Envisioning my hours on Dishes as an exorcism made them endurable.

Three months after I completed my punishment, I moved into Upper School and was nominated Head of the Friends Committee. I had become a well-respected older student. I had paid my dues, been beaten, been broken, and now I was on my way to higher ground. As the head of Friends, I assigned older students (whom I felt were sufficiently dedicated to the school and its teachings) to younger students as Big Brothers or Big Sisters, just as Mira had been assigned to me. The objective was to mold these younger individuals into dedicated members of the Cascade community through love and affection and the constant reassurance that, above all, they would someday love themselves. The Friends Committee was the root of Cascade, the fuel that drove the school; without it, the school would have crumbled.

When I was a child my teachers told my mother that I was a real leader, but that I led people in the wrong direction. Now, as the head of the Friends Committee, I felt I had a chance to redeem myself for all my past negative leadership. I was needed by countless desperate children who latched onto me like suction cups against glass walls. The more inconsolable they were, the more I loved them, the more I wanted to crawl inside their heads and tinker with their brain chemistry, cut, tape, erase, alter—whatever I had to do to understand them and make them happy again.

I had four Little Sisters during my stay at Cascade, but there was one who broke my heart every time I looked at her. Her eyes held the desperation of a gnat trapped between storm windows throwing itself in chaotic circles from one pane of glass to the other. Her name was Lu, and she arrived at Cascade the day after my one-year anniversary. She was only thirteen and disconcertingly thin—not in an anorexic way, but like a

Lara and a few of my Little Sisters

lanky child with a very fast metabolism or a thyroid condition. Her legs were unshaven sticks with bony knees and stretched-out, unmatched socks bunched above her worn Converse All-Stars. She had soft, ivory skin that still retained the purity of preadolescence, and straight, rarely brushed brick-red hair that flowed in every direction when it wasn't knotted into a low ponytail.

She spoke to me in a whisper-like voice, her words barely audible, and I'd struggle to catch one out of every three and quickly piece together the sentence before she stopped, expecting a response. She told me she didn't belong to herself, she belonged to her brother, which confused me until I found out why in a Wednesday Forum. He'd molested her since she was six, but instead of Running Anger at him in Forums, she spent months crying about how sorry she was for him.

Every night during the spring and summer months, Lu and I walked around the pond at sunset and sat on the lifeguard chair at the end of the dock, her thigh half overlapping mine like an eclipse of flesh. We'd lock ankles and swing our legs above the murky pond. Sometimes I'd reprimand her for climbing on the roof of her dorm again, and she'd tell me it was the only place she could think straight. Most of the time we talked about why she felt so sad. I couldn't bear the way she looked at me—like the high beams of a car penetrating your pupils and momentarily blinding you completely. Terror overflowed her eyes when she spoke about her brother, and I felt deeply and entirely helpless. I'd pull her on my lap and start braiding her hair because that was all I could think to do.

Lu loved nature. When she wasn't on Level Three Dishes, she spent her free time planting in the garden or hiking the trails around Cascade. She introduced me to the insects that harmed certain flowers, ones that carried disease or bacteria that would slowly kill them. She knew natural remedies for keeping the bugs away. Aspirin and water. She hated pesticides before it was trendy and frowned on fertilizer, creating her own secret compost pile behind our dorm. She'd sneak down to the garden at 5 AM and mix it in, kneading the dew-soaked dirt with barely fermented vegetables and fruits.

When I left on my second overnight visit, she wrote me this letter:

Dear Al,

Today is my four months. It is about eleven o'clock Friday night and I have your baby pillow right here. It's so soft and comforting. This is the first night I have spent without you in four months and I hate it. Can you listen to Pink Floyd on your visit? Ooopps, I guess that I shouldn't have written that but I'll cop out tomorrow—maybe. See I need your guidance so you better come back soon. I know that was bad, I will cop out. Don't worry. I miss you a lot (even though technically you are still here as I write this) I am sure I'll be missing you while you read it. Al, what happens if I need to cry while you're gone? Never mind, I'll hold it. Besides that is not too likely. Well, I'm going to sign off. Guess what? You're pretty great and yeah, I guess that kind of means I love you lots or something, huh? Take care of yourself out there in that bad world and I'll be waiting for you.

Love, Lu

She hated wearing dresses, so of course Cascade made her dress up every Sunday night, combing her hair and pulling it back with a red headband. Sundays were dressy to begin with; we had a special meal. We wore dresses, held hands, and had to announce what we were grateful for in front of the whole dining room. Lu always wore a pair of brown corduroy pants and a snap-button shirt. Her counselors made her throw the pants out and asked her parents to buy her a dress and have it sent.

The day it came she sat alone by the lake. She held the open box in one hand and fed the ducks that quacked around her with the other. She was crying, wiping her tears with the arms of the dress that she'd twisted into sweaty balls between her hands. She wouldn't speak to me but let me hold her as she cried from the darkest place in her gut. I told her that she couldn't hide behind ugliness, that she had to embrace her beauty and not fear it. I told her that her brother was far away and that he couldn't hurt her here. This was how we talked at Cascade. Still, for weeks she refused to wear the dress, and was forced to scrub the grill for five hours every night until she came around. One Sunday evening, she walked into my dorm with the dress in her hands and asked if I'd help her put it on. She couldn't remember ever wearing a dress before that night.

As the head of Friends Committee, I also had to Run Dishes three nights a week, assign tasks, and monitor the quality of the work done. I wasn't supposed to let anyone go until his or her job sparkled with a shine so brilliant you could see your own reflection. Lu was young and small, and she would always be the last one to finish, usually around ten or eleven o'clock. The grill was a big job, requiring enormous muscle strength and endurance. She had neither. Sometimes I'd pretend that she was do- ing it wrong and grab the heavy Brillo Block from her hands saying, *Let me show you how this is done. I'll never be able to let you off at the rate you're going.* I had to be stern and professional, so the others wouldn't complain. I'd scrub the burnt meat from the rusty peeling surface until spots of tin appeared, all the while instructing her on the task. I'd do it for as long as I could without looking suspicious and then hand the Brillo Block back to her blistered hands. I loathed Running Dishes when Lu was on and of- ten tried to trade shifts with the other committee heads, though she was so often on Dishes that our paths almost always crossed.

Sometimes Lu would leave beautiful watercolors on my bed. Stick figures dancing in grassy fields below a brilliant blue sky and a sun drawn with orange rays dashing out in all directions. There were always two of them, delicate, faceless creatures in blue and pink or orange and green suspended in midair just above the grass. Below them would be a note.

I love you more than anyone in the whole world. You are the most im-portant person in my life. I wish we could take care of each other for-ever. I wish you would never leave me. I hope someday I will be just like you. You are beautiful.

19

THE I AND ME

Fifteen months and eight days had passed since my arrival at Cascade, when I entered the I and Me Workshop. I looked forward to Workshops now; I believed in them. I believed in the tools they gave me, in the realizations they brought forth, in the tears they brought out.

The I and Me:

Inside the Millhouse room many pairs of empty chairs sit facing each other. Below them is a shag carpet the color of a dirty pig's hide. Long rectangular panels of smudged window line the walls. I sit across from an empty chair as I am told to do. A tall, slender counselor with a head of short coiled brown hair zigzags through the empty space between the chairs. She clasps her hands behind her back and walks with purpose, heel, toe, heel, toe. Her name is Karen.

"Welcome to the I and Me Workshop! Please close your eyes and picture a time when you were truly happy. When you felt free and loved. I want you to picture this child. Be there with your child and feel what it felt like." Karen talks softly.

From the stereo "Imagine" by John Lennon plays.

I remember a picture of myself at four years old. I stand on the stone porch of our ranch house in Nevada, wearing white shorts, a navy-and-white-striped shirt, navy Keds sneakers, and my father's cowboy hat. I am smeared with paint and smiling. My right leg is painted into a colorful grid.

John Lennon sings louder now.

The other children around me begin to cry. They sound fake to me, whiny. Cascade students can turn their emotions on and off with the flick of a switch, time to cry—cry, time to be mad—pound the pillow and yell. We are programmed. I am programmed. Sitting in the plastic chair, face in my hands, eyes closed, I am not particularly sad and I hate myself for not being able to cry like I am told to do. I'm not in touch with the pain like I should be. I know I have to go deeper. I know I have to return to where I went in the last Workshop, but it hurt, and I don't want to go back. I hear people walking around the room. I hear things moving or being taken away.

"I want everyone to remember that we still have that innocent person inside of us. The happiness and freedom is still there, too. You just lost it for a little while. This is your Me. This is what you are here to fight for. Now everyone please open your eyes," Karen says.

In the chair in front of me sits a framed picture of a little girl at four or five years old. It's me, and I am holding my favorite doll. My hair is brown with blond summer highlights and pulled to one side with a quilted barrette. I wear a white blouse with blue smocking and yellow ruffled shorts. I am smiling with a mouthful of baby teeth, white and evenly spaced on both my upper and lower jaw, and my eyes are squinting against the sun. I look at the girl in the picture and think about the girl who stares back at me from the mirror every morning. I don't believe they are the same person. I don't believe the little girl still exists inside me. I think Karen is full of shit, and Michael Stipe is now singing. He is telling us to hang on, hang on, hang on, even when we think we've had enough of this life.

It's the music that drives me to tears. It is always the music. I don't like REM or Michael Stipe, but he just sounds so sad singing those words, like it's such a struggle to simply breathe, and he keeps telling me I'm not alone and that we all hurt, and I feel like he is talking to me. Like he wrote the song for me. And again he tells me to hold on, and I look hard at the little face in the picture with the button nose. I look for answers. But she just giggles back at me, frozen forever in mid-run.

Sam has fallen off his chair and is curled into himself, hysterically crying. Karen rubs his back, and her cheeks flush a hopeless red because she knows he doesn't remember a childhood, and she can't fix that. Michael Stipe sings: *No, no, no, you're not alone!*

Another counselor comes to the center of the room. It is Katherine; I can sense her even with my eyes closed. She is petite with dark leathery skin and some wrinkles. Her hair is chestnut brown, short and wavy, barely brushing her shoulders. She is my Upper School counselor, and she is the woman who held me after my first visit with my parents, after I'd been at Cascade for only three months. I like Katherine. She is strict, but real and fair. When Katherine speaks, I listen:

"Now imagine this child in front of you playing, doing whatever you did as a child, whatever fulfilled you or made you whole. Picture the activity and be there with your child!" She speaks loudly and paces slowly around us.

I see myself in Connecticut, thin and light-boned in shorts and blue leather sandals, alone in the woods. I carry sticks to make teepees, place plastic hot dogs inside old Indian ovens made of stone. I talk to the plants that I grow and name them and feed them like children at dinnertime. My little hands, encrusted with dirt, lift and pile and rake as I build my untouchable world.

"Now think about what you've done to this child, what you have taught this child, what you tell this child every day! And I want you to tell the child all these things!" Katherine sounds angry, and the room is silent except for the whirring fear and Michael Stipe, who is fading out.

"If you don't understand, I can give you an example! Who would like to volunteer?"

She walks by me, then pivots on one toe and returns. I am shaking with dread. She falls to her knees and her knobby hands clasp the sides of my thick glass picture frame. Her mouth opens—too wide—like a marionette's, and she pulls the picture close to her face. She begins to scream. My little face is still smiling as the glass begins to fog with her breath.

"You're a failure to everyone! You'll never succeed! You drove your mother into her depression! It's your fault she's an alcoholic! You're going to kill your own father! You're a drug addict, a bum, part of the lowest life form, and that's all you'll ever be! And you'll always be alone, Alison! No one could ever love you!"

I am crying, and I feel Katherine's hand on my back as she tells everyone to start. I am hysterical because what she was yelling is correct. That is how I feel, and that is what I tell that little girl. All I can hear is screaming and crying. The speakers blast another song.

I can't do anything but cry as Katherine yells and yells. She wants me to yell at the picture because that is what I do to the little girl every day. And I feel so much inside myself, but I can't say a word or move my mouth or even consider putting a sentence together. I hear counselors antagonizing other students. I listen to them and try to block Katherine out, but it is impossible, and finally I crack and begin to scream at the picture. I tell the little girl that she is a failure, that she wasted her life, that she is an awful person who deserves to be trapped in this nuthouse. I tell her she is a drug addict who will never amount to anything. I tell her nobody will ever love her, that nobody will ever be proud of her, that she will always be alone. I believe it, every word of it.

I am in a parallel world where only voice exists, where my face invades my body and I scream into myself. Sweat, tears, and snot pour onto the picture frame that I hold tightly, close to my face, and they merge with each other like watery eggs and potato slop at a diner.

Again, I feel Katherine's hand on my back. She is trying to bring me back into the room. Slowly, I begin to be aware of myself. I know I am present when I lick the wet of my chewed lips. They are jagged and raw. White skin lifting itself like air bubbles under stickers you want to smooth.

"Now switch! Get in the other chair and yell only positive things at the child. The things you know somewhere deep inside yourself to be true. The things that the little child in that picture believed," Katherine says. "This is the Me fighting the I, and you must destroy the I, the negative voice lurking in your head *all the fucking time*!"

I rise, still crying, and place the photograph on the other chair. Now I am meant to embrace my Me, the positive, confident person hiding inside myself.

"Go! Louder! Come on!" Katherine screams, cheering for us as she wanders the room.

More counselors appear, five, then ten, walking in circles around the room clapping and cheering and saying fight, fight, fight. The *Rocky* theme song comes on the stereo: "Eye of the Tiger."

"C'mon, Alison! Fight, fight for yourself!" Katherine yells an inch from my face.

At first I don't believe it will work. I don't believe this will fix me. But as she yells and yells, I decide it can't hurt. It can't make it any worse. What do I have to lose? I break and yell and fight and yell like I've never yelled before. Like my fucking sanity depends on this moment, on these words. Like this exercise will save me!

I am completely immersed in this workshop. I've allowed myself to travel deep into the recesses of my mind. The bones in my skull create spaces that no one else can see, but I swim and dive and slither inside. If sadness has a definition, it grows inside of me, expands like sponge animals in water, filling my cells. I am hysterical with regret and nausea and loneliness. I lick my upper lip and taste metallic blood. My nose is bleeding and I wipe it with my arm.

The cries dissolve, and we are told to mingle. We walk around the room and stare into each other's washed-out, hazy eyes. Our eyes are the windows into our souls, and, if we look through them, we can see all that is pure and good. That's what the counselors tell us. And we believe them.

"You have completed your second-to-last Cascade Workshop.

As a little girl, on the Lazy W Ranch in Dad's
cowboy hat, painting. This is the memory I
returned to in the "I and Me" Workshop.

You should be very proud of yourselves. Let us all link arms and return home to the other Families," Karen says.

We link arms and form a long chain like a succession of streamers at a party. We walk up the hill, along the pond, over the bridge,
and into the Cascade House, where one hundred thirty other kids
wait to welcome us back from our Workshop. They are standing in
a swaying horseshoe with arms around each other. "I Can See
Clearly Now" by Johnny Nash is playing on the stereo. Some
younger students bite their nails while stringy bleached hair hangs
over their faces, others refuse to stand in the circle and line the periphery, but most stand teary-eyed and weary with heads resting on
shoulders or arms wrapping waists. When we enter, they all come
out to hug us and congratulate us, and we smile and say thank you
and cry a little more.

20.

HOME VISIT

I was sent on my first home visit on March 28, 1995, sixteen months after my arrival at Cascade. From the airplane New York resembled a blinking silver chess set. The positions were different, but the pieces looked the same, some curved and short, others tall and rectangular with rigid, crown-like tops. Coming from the natural environment of Cascade, the city looked like a cyber-industrial backdrop for a movie set involving cloned replicas of human beings. I didn't feel that warm tingling in my chest one should feel upon arriving home after a long hiatus. I felt just the opposite; an arctic chill seemed to leak through the airtight windows of the plane, filling my body with an anxious shiver of impending doom. The one saving grace was the sunshine that softened the edges of the buildings with granular gold outlines, but as we hit the runway, even that disappeared.

Someone was snoring in the row next to me. The stewardess came on the loud speaker, *It's a lovely clear night in the Big Apple. We should be at our gate momentarily, folks.* I packed up my Walkman. Inside it was a tape of Cascade hits that we'd been given last Christmas. I suddenly realized how embarrassing it would have been if someone had asked me what I was listening to, and I popped open the cassette, stuffing the tape in the zippered compartment of my carry-on bag.

My mother's feet were the first recognizable thing to come into sight as I rode the escalator down toward the baggage claim of JFK Airport. Long and unusually large, dressed in nude stockings and slipped inside a pair of fashionable loafers, they peeked out from below the hem of her cashmere wrap.

"Hello, dear," she said, reaching for me.

"Hi, Mom," I said.

"Daddy is in the car. It's really nippy here. It was in the single digits last night." She tried to help me with my bag, but it was too heavy for her.

"I have it," I said.

"Is it cold at the school?" she asked.

"I didn't notice," I replied.

The large glass doors slid open behind us and we turned toward the exit, making our way into the cold New York night. Sirens and horns and the whir of plane engines buzzed around us. A delicious whiff of cigarette smoke blew my way. On the other side of the platform, I could see my father reading in the backseat of a gray car, and as we neared it, my mother pointed limply with her ringed hand, saying something that was muffled by the wind. Her long shawl trailed clouds of a familiar perfume as she walked.

Once we were on the highway, she began to question me on every infinitesimal detail of the flight, filling the heavy car air with meaningless chatter, as if she were just like the other mothers piled in baggage claim welcoming their children home for spring break. The driver switched between lanes with jerky movements, and my father remained silent, seemingly enthralled in his biography of Churchill.

I recalled the last time I had been with my parents in an Esquire car driving from JFK Airport to our apartment. It was the January before I was sent to Cascade, and we were returning from the Lyford Cay Club in Nassau, Bahamas. I sat sandwiched between my mother and father with my nose in some trashy teenage novel. A concentrated aching silence hung in the air between us, a familiar feeling our three bodies created when forced together during those years. Mother stared vacantly out the car window clutching her Hermes Kelly bag as yellow headlights washed across her face. My father sat to my right with his briefcase in his lap and his chin resting in his palm. We were in traffic on the FDR, and as we passed the Seventy-second Street exit a short spurt of gas entered the car.

The driver turned around, staring at us with a deplorable expression of shock. My mother slowly rotated her head from the window and said, *What was that?*, attempting to pretend it was something other than what it obviously was. *Dad farted,* I said, proudly smiling up at my father who was struggling to suppress the childish grin materializing across his face. And then, as if giving in to our true nature, we both began to laugh hysterically as my mother continued to express her horror and revulsion at our barbaric behavior by saying, *That is not funny. Would you two stop egging each other on? You're acting like children.*

Pulling up in front of the familiar green awning of my apartment building, we were greeted by two doormen in English caps rushing toward the car, one opening the doors and the other unloading the bags from the popped trunk.

"Welcome home, Miss Weaver. We missed you," Jesús said, nodding.

"I'm not home for long," I said.

"Oh, more school then?" he said, blowing out a breath of relief.

When I entered my bedroom, I discovered all my belongings were gone; they had been bagged and twist-tied and sent off to Goodwill, per Cascade's instructions. The space was as bare and destitute as a hotel room, with only necessities at hand. In the bathroom, two towels were folded and hung over the silver rack; a glass tissue box and a water glass rested on the marble counter. Even the cabinets were empty, my makeup gone, my lotions, perfumes, and toothbrush. A single bottle of Advil sat alone on the first shelf, presumably for the occasional guests who now occupied my room. It was as if I had never existed.

I sat on my windowsill for the next hour and watched the people of Sutton Place. I saw men in suits, slightly wrinkled from a day at the office, arrive home and disappear under green or black awnings. I watched hands in white gloves open doors and men vanish into gargantuan brownstones. I imagined them pouring themselves tall glasses of eighteen-year-old Glenlivet, slipping their tired feet into some cuddly deerskin slippers purchased on their ski trip out West last year, and settling down for the evening news. I imagined their wives lying in bed flipping through fashion magazines, instructing the cooks on the evening meal or curling their hair for a benefit they would attend later that night. Maids walked dogs in the cul-de-sac below me, and nannies pushed strollers or held impressionable hands, while sharing neighborhood gossip. The anemic trees lining the sidewalks were still lit up with Christmas lights and twinkled on and off in different beats.

Behind a window in the townhouse across the street from me, a boy played Nintendo. I recognized the game and watched the bright characters scamper across the screen jumping when the boy hit button A, rolling when he hit button B. Within minutes the boy had completed the game and threw the control at the television set. Glimpsing me, he pulled down the shade halfway, paused to flick his middle finger in my direction, and pulled the shade to the floor.

"Ali, dinner ready," Mai said, knocking on my door.

Mai was Filipino—a small, kind woman who hummed show tunes as she cleaned and called my mother *Ma'am*. Mother made her wear a gray and white uniform with a scalloped collar, and I thought she looked like a little fairy flitting around the house with her dusting tool. She'd been with us for four years. I thanked her and walked to the dining room.

Dinner was long and painful. We sat spaced evenly around the oval table. Mai moved quickly and gracefully from one person to the next, holding a hot platter of food. I watched my mother move the food around her plate. Green beans with crushed hazelnuts sat like logs one on top of the other, and she removed them almost as carefully as one might remove a pick-up stick from a pile. She flattened the mountain of pureed orange sweet potatoes into a thin layer. Then she cut the lamb chop into pieces and pushed them under some green beans. I had learned early how to abstain from taking a single bite of food while making it appear to the amateur eye as if I'd politely tried everything. This was one thing my mother taught me that was beneficial in my later years of vegetarianism. It wasn't that she was anorexic; it was simply that her stomach was constantly filled with alcohol. Often, though, she would crave salt late at night and order a pizza from around the corner. Growing up, our freezer was always filled with ten or eleven Saran-wrapped slices of leftover pizza.

That night, I observed my mother sipping her wine and playing with her food and dabbing the sides of her mouth—sip, play, dab, sip, play,

dab. Every movement so premeditated. The faint flicker of candles patterned her face with shadows, and even the dim light of a candlelit evening could not disguise the reddening blood vessels around her eyes. Yet
she shifted in slow movements, trying hard to look dignified.

Past the French doors at the end of the long mirrored living room,
classical music bellowed from the custom-made speakers. My father sat
silently at the head of the table working on his salad course and conduct-
ing in between bites.

"It's Bach. Prelude number two," he said, aware of my stare.

"It's loud," I said.

"That's the only way to listen to it, Alison," he said. "You really can't
feel it otherwise."

"Oh," I said.

I wanted nothing more than to leave the apartment and catch the next
plane back to Cascade. It seemed my only salvation, a place so opposite
the coldness in that dining room. Bach's prelude was punctuated by the
clinking of silverware against china and the suction-like noise of chewing.

The thick silence hanging between our three bodies hurt to sit in.

"Tell us about school, dear," my mother said.

"It's not a school," I said.

"Don't you go to school during the day? They told us that was why
the place was so damn expensive," my father said.

"Alison, you know you go to school there," she insisted.

"Yes, I go to school there," I gave in.

"Can I get more mashed potatoes and another scotch, Rose?" he asked.

Mother rang the silver bell, and Mai came in and took his order.

"Where are your manners, Ali? We don't sit with elbows on the table,"
my mother lectured.

"Actually, Mom, we crazy people are allowed to sit however we like
back at the nuthouse. Manners aren't really a priority," I rebutted.

"Could you please make an effort to get along with us?" she said.

"I'm finished. Can I be excused?" I asked, pushing my chair back.

Neither of them responded, I took their silence as an invitation to do
as I pleased, tossed my napkin onto my plate, and exited the dining room.

Later that night as I got ready for bed, I found a Lu watercolor with dried flowers taped to the paper and a note inside my duffel bag:

Hi, it's me. See the picture of the two little girls on the front? That is us. You know who is who and that is how I want it to always be, us looking out for each other and holding each other when we are sad or scared or hurt. Al, you are so very special to me. I hope you are all right on your visit. I am not all right here without you. Will you write me a poem out there? I love your poems. I'll write you one. Come home soon!

My father's hard knock came from the door.

"Come in," I said.

"You're okay?" he asked.

"Sure," I said, my cheeks flushed, eyes averted from his glance.

"Glad to have you here," he said, awkwardly.

"Thanks," I said. "It's weird to be back."

He nodded, his hand pulling the door languidly behind him as he cleared his throat in the hallway, and the brass wedge gradually pushed its way into place with the familiar ticktock. I could feel his mind frantically searching for something to say, a comforting anecdote, an inappropriate joke—anything. From across the room, I waited, confident in his response, knowing my father was never at a loss for words. But that night, all I was met with was silence.

When did it end between us? When did we lose that seemingly indestructible adoration we'd once shared for each other? I suppose love gets chipped away at over time, like anything else. I can certainly remember moments when I let my father down. I can remember moments when he looked at me with eyes that no longer recognized me, moments when I looked back with contempt.

I can still remember what his eyes looked like the first time I failed him. We had taken a trip to the hardware store in Connecticut. We'd purchased chickens on the way home. We'd spent the afternoon deep in the recesses of our old shed, behind tractors and mowers and tools, designating an area to build them a home. He'd spent hours designing the cage, drawing up architecturally sound plans with rulers and markers. He'd measured and designed and measured some more, as I stood around picking splinters from the splitting beams.

In an hour we had the frame of the chicken coop up, and by dusk it was built, a rectangular-shaped cage held together by wooden beams and covered in mesh wire.

The next day we were sitting in the car on the way back from the library, my small hand fiddling with a pen I had hidden in my jacket pocket.

"What do you have there in your pocket?" My father asked, noticing it.

"Nothing," I said, pulling my hand from the warmth of the pocket.

The pen tumbled onto the car seat. It was purple with pink swirls and glimmering silver tassels that turned the white light on the windshield into shimmering rainbow spectrums.

"What is this?" he said lifting the pen and shaking it in my face.

"A pen," I said.

"Yes, I know that. Where did you get it?"

"I don't know."

"Alison, don't lie to me," he lifted the pen to his eyes, reading *Easton Town Library*.

"I found it in the library," I said.

"You stole this from the library?" he asked.

"No!"

"Yes, you did! I can't believe you would steal. Why? Why did you take this?" he asked, shocked, his mouth wide open.

In his eyes was something I hadn't seen before, something dark and disappointed, and I didn't know how to fix it, how to make the light and pride return. My father's idyllic view of his little girl was whipped out the open window like ash from a cigarette that day. He drove me to the library after our errands, and I apologized for stealing the pen between sobs and coughs, but when we got home that afternoon, he didn't take me to the barn to play with our chickens, instead coldly instructing me to go inside and practice my multiplication tables.

* * *

After my parents had gone to bed, I poured myself a tall glass of milk and wandered through the apartment. I walked into Ilse's old room and sat Indian-style on her bed. After five years I could still smell her there, on the sheets and in the closet where her abandoned white uniforms hung stiffly on hangers like a procession of cutouts. My entire childhood felt contained in this woman's eight-by-ten-foot room. I suppose I was looking for something that held even the tiniest bit of comfort, and her room was the closest I was going to come.

I decided to take a walking tour of the apartment in hopes of piecing together my life, which had, over the past few months, evolved into a complex, intimidating entity. Details I'd once thought incidental or extraneous had begun to appear vital, and like weeds parting to reveal the earth they grew from, my life was beginning to make sense.

Our laundry room smelled of Downy as it always had, and in a straw basket on the floor were monogrammed bedsheets and towels, ironed, folded, and ready to be put away in various linen closets throughout the apartment. On the dryer were piles of my father's shirts, underwear, and socks, all neatly stacked in line to be ironed.

I unlocked the back door and took a seat on a cold gray stair. I tried to remember the last time I had been in that stairwell, and a murky image of schoolgirls in plaid uniforms appeared. It was us, my old friends and I standing in groups of three or four, thrusting our chests forward as we practiced French inhales or smoke rings. I could see Niccola looking exotic, with her silky black hair falling like a cloak seductively around her, blowing the best smoke rings of us all. Samantha looking around idly with expressionless eyes, wondering when the next Gucci sample sale was, and Lauren, my best friend, on my lap as usual, pretending to inhale her smoke as it came out in cloudy balls.

But I didn't go into the back stairwell for those memories. I went there in search of an answer to the question that had come up during my first workshop at Cascade. I would never know if the Australian man had raped me. It was Marlene's word against mine on that one, but I needed to know if *Elliot* had really raped me. When I saw him in my memory, standing in front of me with his long-soiled hair, shaggy, waist-hung corduroys, and piranha-eaten Patagonia, twisting his pinky ring like he always did when he didn't want to speak, I didn't hate him. I didn't picture him as a monster. How could I? I had slept with him again and again and again. Elliot Windsor III.

Elliot loved sinks, loved to spread me open on the edge of one and jam himself into me. He loved the smell of tree bark and sap, loved to mold my naked body against huge oaks, angling my shoulder blades until they formed an ivory butterfly ready for flight, then he'd grab the trunk and me in one gulp. Cold basement floors were good, dusty enough to prove he wasn't a sissy, so that when he got back to the bar with a soiled polo shirt, his friends would know where he had been. Cars, roofs, locker rooms. Beds were not his forte.

The night it happened we weren't in my basement stairwell as we had been many times before. We were in the stairwell of a clubhouse in Tuxedo Park. I lay flat on my stomach against the stairs, my chin resting on the edge, my chest caving from his force. He liked me at an angle. I know I screamed uncontrollably when the pain shot through my body, up my spine, vertebra by vertebra, as if a cannon had been set off inside me. It was after 1 AM on a Saturday night, the members of the club were all gone, and Elliot didn't tell me to be quiet but placed his fingers inside my mouth so I'd bite instead of scream. Afterward, I poured the leftover wine from the bottom of the empty bottles to wash away the blood and picked up our cigarette butts. He kissed me goodnight and rang the back elevator.

Marlene had insisted it was rape, that my noncommittal ambivalence was a protection device so I wouldn't have to confront the trauma. I wasn't sure. Elliot and I had had sex before, and he'd never tried it that way, but that night he plodded in without restraint, ramming me from behind again and again as if my shrieks of pain were nothing but the euphonic soundtrack to a grade B horror film. A few weeks later we'd have sex again, and this time he'd rub me against green asphalt until the skin covering my spinal column bled, but I was drunk or on pills, and it didn't hurt at the time, though later I remembered tears falling from my eyes and him wiping them away.

I wasn't screaming *no, stop, you're hurting me*. I wasn't kicking or wiggling from his grip. I was crying silently, biting his fingers. A decent human being, when noticing my discomfort, would have stopped. But I always knew Elliot wasn't a decent human being. That was part of the attraction. What Elliot didn't realize was that I allowed these interactions to happen because I didn't believe I deserved tenderness. I didn't think I was worthy of anything softer, anything more genuine or pure than whatever rough, haphazardly placed caresses he was willing to dole out. I never thought to question it. And I suppose my acceptance of his brutality gave him the sanction to move forward as he wished without guilt or remorse. So then who was I to claim rape years later?

My parents' bar was located in the library; it was a boxy, cluttered room tagged on at the end of living room. This was the room where they spent most of their time. The room they'd sit in during the evening while they watched the nightly news. The one in which they would eventually eat dinner because they stopped liking each other enough to have civilized conversation, and the TV was a good distraction. It was the room where my mother sat until 1 or 2 AM sipping wine as puffs of smoke circled her like lost balloons, and the muted television flicked fragments of light across the bookcases. It would be the room where my ninety-one-year-old father would spend his final days, lying diagonally across the sofa, the napkin tucked into his polo shirt forming a patterned diamond across his sweater.

The room always seemed illuminated by a hazy, reddish glow and usually smelled like citrus. On the far wall was a beige and red sofa with matching pillows trimmed in silk tassels. Above the sofa was a primeval mercury mirror speckled with deterioration, bought at a Sotheby's auction years ago. Across from the mirror was a marble fireplace with French armchairs on either side, and Mark Hampton tables painted with a finish meant to evoke texture and age. Leather-bound books lined most of the room's bookcases, interrupted only by an occasional picture: my father shaking the hand of Richard Nixon, my father dancing with Ginger Rogers, my father and his godson with the Queen of England, a group of men in suits, a family photograph taken by the canes in the foyer. The walls were a dark wood paneling with high moldings imported from England; the room had, in fact, been brought over from England by the previous owner. My mother called this room *a classic and traditional Upper East Side library.*

My favorite chair in the room was the antique English guard chair, its edges outlined with gold upholstery tacks and a wooden compartment below its seat. It was said to be placed outside castles for guards to sit on back in the 1700s. Their lunches and bedpans would be stored in the wooden compartment so they never had to leave their post and subject the castle to danger. When I was a child, I used to sit in that chair and pretend I was a guard. I'd sit straight as a board and nod at imaginary people saying *Good day, Ma'am* or *Good day, Sir*, and after a while I'd lean over and open the compartment and begin to eat my imaginary lunch.

The bar was a tiny room in the corner secured by a gleaming wooden door with a circular gold latch. And as I made my way into the library that evening, a desire to drink gusted through me. Everything in the room was so dustless, so sparkling, so uninhabited, and the perfection of it all saddened me because I knew the truth. I knew what lay behind the happy appearance, and I needed something to kill that knowledge.

The door to the bar wasn't locked. I opened it slowly, trying to minimize the drawn-out creak the hinges had made since I was a child. I could smell the lemon and olive juice ingrained in the cracks of the wooden cutting board. I could smell my father's watered-down scotch glasses sitting next to the ice bucket. On the bar door hung a photograph of him, taken by Yosuf Karsh, that reflected onto the mirror behind the glasses and sprawled out in every direction. In the photograph he wore a red jacket bearing the numerous Medals of Honor he'd won during World War II. His eighty-year-old hands were folded neatly in his lap, his smile proud and vigilant, narcissistic.

I remembered the day the picture was taken, the man's studio, walls and walls covered in his clients: Winston Churchill, Prince Charles, the Queen of England, many famous, many dead. My father wanted a picture of us taken as well, even though my lip was pierced by a silver ring and my eyes smeared with green eye shadow. Karsh said he could airbrush my face.

When the negatives came in the mail, my mother said none were flattering of me and insisted we not waste three hundred dollars on a print we would never frame. She was right; I looked awful. My bleached white hair was brittle and stringy, its shards falling from my head in uneven clumps, my smile was crooked, and my blue nails were bitten and chipped. But my father ordered one anyway.

I stood at the bar staring at the array of bottles before me, tall shimmering things with gold or silver collars standing like soldiers at attention, and after a minute amount of deliberation, I plucked a clear bottle of Absolut from the back row and reached for a glass. The lucent liquid poured out of the bottle top with ease and grace, settling around the three cubes of ice I'd managed to save from the half-melted remnants in the bucket. The cubes clicked and squirmed uncomfortably. Behind me I could see the room in all its flawless splendor reflected in the mirror with my many fathers, and then in a strange moment of clarity, I realized that sitting in that room and drinking would be capitulating to my parents' world. It would be saying *me too, take me too*. And I would become nothing more than another disillusioned character in the lonely story. I put the bottle back and closed the bar door behind me.

It still perplexes me that I didn't drink at all during that visit. But as hard as it may be to imagine, the decision, when contemplated for a few minutes, was easy to reach. I knew I didn't want to surrender to that world. And I also knew I wanted what Cascade ostensibly had to offer me. Drinking would have prohibited that. Lying or withholding information was by that point in my stay out of the question. I was plugged into the rhythm of the school. Their guilt was my guilt. My shortcomings, my failure to follow the rules affected everyone else. I was their example, their leader. I couldn't fail them. I couldn't fail the person I had become to them. Why I didn't drink that night had nothing to do with actual recovery or self-love and everything to do with the fear of losing the place where, for the first time in my life, I felt loved. I knew if I drank, this love would be revoked, and I'd be demoted, lose my committee head, all my privileges, my little sisters. The decision seemed simple then. I had built a life for myself at Cascade, and I wasn't willing to let it go.

21.

THE CONVERT RETURNS

I returned happily to Cascade after my home visit. Cat Stevens was once again playing on the stereo, and the House smelled of cedar and mothballs, a familiar and comforting scent. Lu ran toward me, her deep red hair shining in the dim light of the room, tumbling in curly waves down either side of her neck. It was Sunday, so she wore a flowing paisley skirt, a white blouse, and childish schoolgirl sandals that matched the red of her hair. I looked for Mira. She was in the corner of the room holding a hysterical younger student in her arms. Sam had returned from his visit a day before me and admitted to drinking. He'd been put on All Day Dishes and Bans with the entire school. It broke my heart to see him moping around the campus, forbidden to make so much as eye contact with a single person other than the counselors and his Big Brother. That night he sat in the corner of the House drawing comics and graffiti tags on his sketch pad, looking up to smile when he noticed me, but quickly returning his glance toward the paper in front of him, fearing someone might notice.

Students were gathering in the main room for Last Light, a fifteen-minute period during which we all gathered together before bed. Announcements were made and sometimes people sang songs or did skits. If a tragedy had occurred in the real world, we would hear about it during Last Light and have a brief group discussion. We heard about the Oklahoma City Bombing during Last Light, the deaths of Timothy Leary and

Ginger Rogers and Jerry Garcia, but the worst announcement during my two years was the April 5th suicide of Kurt Cobain. Students became hysterical, and we gathered in circles to sing his songs, even though they were unacceptable. One girl's younger sister killed herself the next day, leaving a suicide note to Kurt; some students even Ran Anger or cried about him in Forums. He was our god. A man synonymous with the disillusioned and unhappy youth of the nineties. He read our minds, knew our problems, and spoke our language when no one else in the world did, and now that he was dead, many questioned their reasons to continue.

The school evolved into a state of mayhem for weeks; most younger students refused to follow the rules, fifteen ran away, some broke into the nurse's station and looted it for prescription pills. There were only about twenty of us who were unconditionally dedicated to the program, and it was impossible for us to control the insurgency of the other hundred and thirty mutinous students. Eventually, the head of the school, Taylor Gilbert, was called in. He canceled all classes, therapy, and Forums, and after a frighteningly dictator-like speech about the state of our school, we were put to work. Mr. Gilbert was a terrifying man with bleached blond hair and leathery, brittle skin, the result of his permanent suntan. He wore light blue jeans, penny loafers (with the penny tucked in the front), polo shirts, and navy blazers, and rarely was he seen among us for more than an hour every few weeks. For days after his state of the school address we did nothing but scrub and dig and chop in silence until at last he gathered us in the House and asked if we were ready to commit to Cascade and to ourselves. Things quickly returned to normal, and Kurt Cobain's death became a thing of the past.

In February, Mira graduated and moved back to New York City. She gave me a brown suede journal filled with self-empowering quotations, sentimental notes, and pictures of us taped throughout.

"The second half is for you to write in, and I'll read it when I come up and visit so it won't really be like I'm missing out on anything," she said, as we stood on the steps of our dorm in the still-dark morning.

"When will you visit?" I asked, tears streaming both our faces.

"As long as I remain sober for two solid months, I can come up any-time after that, so maybe for your birthday in April," she said.

We were unable to let go of each other. In the distant parking lot a van shot spurts of diesel fuel into the dry air. It contained the seven students being taken to the airport with Mira. We held each other, our bodies shaking and hysterical, our mouths sucking back breath, our heads nuzzled into the napes of each other's necks. I think, subconsciously, we both knew that things would be different out there, that what we had at Cascade might not survive. Finally, the driver beeped his horn and yelled some indistinguishable words.

"I love you, Alison Weaver," Mira said, "and remember, if you need anything at all, just write me, and I promise to write right back."

"I will. I love you too, and if you need anything you write me."

"Okay," she said, letting go. "I've got to go."

"I know," I said.

She waved to me from the backseat as the van slowly turned out of the parking lot and down the dirt road. It was still dark out, but a layer of deep orange lined the jagged mountain tips like icing on the top of a chocolate Halloween cake. The temperature was well below zero, and I wore two fleece jackets, a down parka, a hat, long underwear, and my pajamas, and even though I could feel my feet numbing, I couldn't go inside. I had to be brave for the younger students. I had to be brave for Lu. They couldn't see me like this, hysterical and desperate. My friendship with Mira had kept me alive during my first year at Cascade; it was hope and salvation and a reason to wake up in the morning. She had taught me to believe in their teachings and follow the path they laid out for us, and when I questioned it, she reminded me of the bigger purpose. Now I was a convert, a believer in all that Cascade offered, but living there without her still frightened me. I had other friends at Cascade, but none I trusted like Mira.

I sat alone on the wooden bench by the duck pond reading the journal fervently from front to back, hoping it would ease the hollowness her departure had left. I could barely make out the words through the tears falling from my eyes. Some were poems written in a black calligrapher's pen filled with lots of loops and swirls; other pages were huge colorful murals with Mira's tag name—Scandal—graffitied at the bottom. As I read, my fingers began to thaw, and the tip of the sun peeked out from behind the mountains. Day was starting, I could feel it. A bustling noise had developed within the wooden cabins—students were awake staring at ceilings, marking their calendars with big black X's that counted their remaining days at Cascade, scrubbing toilets, or evenly spacing the hangers of their closet. I took a deep breath, steadying myself, put Mira's journal in the pocket of my down coat, and looking out at the lake for a moment more, resolved to go on without her.

My second and final summer at Cascade was hot and dry even for northern California. The dirt paths that zigzagged between the buildings were cracked and hardened, and all signs of life had vanished. Weeds, ants, snakes, even the hyperactive lizards had all left for cooler air. My white socks remained perpetually brown, leaving piles of dust outside the shower when I removed them at the end of the day. It was futile to wash or shampoo your hair because within five minutes of being outside you'd be covered in dust again, but we did it anyway. It was a rule.

I spent my free time swimming in the muddy pond, and eventually grew accustomed to the leeches that would suck on my legs and the water snakes that sometimes slithered next to my waist as I swam. The hot, dusty land sucked life out of me as I breathed, but the water replenished it. I loved the dark pond undulating against my skin as I propelled myself along, my breasts skimming the murky bottom and causing dirt to rise in clouds around me.

Lu and I would often have diving competitions or races from the dock. There was one particularly cloudy spot on the surface of the pond she refused to swim over unless riding my back, and this soon became her favorite way to get around the pond. She'd squeal into my ear, nearly deafening me in her state of half-glee, half-terror.

Sam and Jake were both off Dishes and Bans, and spent their summer running shirtless through the woods with sticks in their hands and black-berry juice smeared across their chests and faces in some sort of tribal de-sign. They pretended they were chiefs of two different Indian tribes. Sam came alive that summer. He seemed to have forgotten about Jerry. No-body spoke about the incident anymore, and I imagine the school wanted it to stay that way. He wore his baggy jeans and skater shoes, and no one uttered a word of criticism in his direction.

In August, I embarked on a sixteen-day hiking trip through the Trin-ity Mountains with fourteen younger students and two counselors. It was an honor to be awarded this privilege, and I took on the role of assistant counselor with fervor. Cascade was the first place I'd ever felt dignity and respect from adults, and I reveled in it. When you've been disparaged and criticized your entire life, it is most gratifying to finally feel that you're being recognized as something worthwhile, even if it is only in a micro-cosmic Orwellian mountain community. We take what we can get.

My title was Trek Support, and I was grouped with two other Trek Support students, Mike Preston and Marilyn. Mike was a tall, thin, soft-spoken young man with pudgy flushed cheeks and a slightly contorted nose that had been smashed with a baseball bat during a Little League game when he was nine. The bone protruded off at an angle and one side seemed slightly swollen if you stared at it long enough, though the re-maining flesh formed a cute button speckled with freckles. His eyes were wide and hazel, and his hair fell in dark black waves around his face. Though he was a member of my Family and I had been through seven Workshops with him, we had never exchanged more than a few friendly words here and there.

My days on Trek were spent on eight- to ten-hour hikes that switch-backed up and down the Trinity Mountains. Marilyn would usually lead the group in a Beatles sing-a-long as we climbed the mountain spurting lyrics between gasps for breath. We'd stop for lunch by various bodies of water: streams, lakes, waterfalls, and we'd jump from rocky cliffs into the glass-colored water, or slide down tame waterfalls that never failed to turn us black and blue. My hair was braided into thirty half-inch braids, and everyday I wore a white V-neck T-shirt and blue shorts with "Summer

Mike and me on Trek

Trek 1995" written in bleach across the front. At night, we washed our clothes in the creeks and remained in our bathing suits until darkness fell, and the warmth of the day was snatched away, sometimes sending the temperatures as low as twenty degrees. Students gathered around the campfire, setting up stoves, pouring powdered hummus into water, or cooking beans and rice. It was a simple and pure existence; everything was wrapped in a selfless decency that everyday life could never have come close to touching.

It was on Trek that my one significant Cascade romance occurred. I remember the day I first took notice of Mike. It was during that ambiguous stretch in the evening that can't really be categorized as day or night when light transforms to darkness in an anonymous moment. I squatted by the stream, rinsing out my clothes, rubbing them between rocks to remove the caked dirt and perspiration, when Mike came up beside me. I can't recall what he said or why we began conversing, just that I teased him about something, and he pretended to push me into the water, a silly high school flirtation, but I liked him immediately. I liked him for his innocence, his softness. I liked him because I knew he would never touch me the way all the other boys of my youth had.

That night Mike and I sat by the campfire long after everyone had gone to sleep. We exchanged stories; he had been addicted to PCP. He'd lost his best friend to leukemia at twelve years old. He told me he wanted to be racecar driver, and I laughed. We didn't mean for anything to "happen," but two eighteen-year-olds left alone in the middle of the woods baring their souls to each other are bound to feel a surge of sexual energy. It started with his hand on my thigh, my hand on his arm, our breaths mingling with the smoke of the fire. I tried to resist. I felt the guilt all around us, but I was enraptured by the moment, the trees, the fire, the stillness of the night. I wanted to crawl inside him forever. Safety seemed to exist inside of him, nowhere else in the world but inside this boy.

Eventually we were in the sleeping bag and at each other like cats in heat. His touch was gentle and warm. He rubbed my thigh and kissed my neck. Each stroke between my thighs grew longer and slower, rising higher up my leg. I panted into his ear, hardly able to contain myself, quivering under his touch. His hand eased up my leg further and further, a body immersing itself in frozen water, ankle, knee, waist until completely submerged. Finally, Mike glided his hand gently between my legs pushing my flannel pajama pants into the creases below. I could feel him against my leg, hard and stiff, and I mustered up the courage to feel him, to explore his shape and size and trace the outline of him against my thigh. His breath smelled sweet and hot against my cheek. We investigated the folds of each other's bodies. His hands on my breasts, against my skin and around my neck were comforting; I felt safe in the arms of a man for the first time in my life.

But as we lay there, holding each other, I could feel guilt rising. I could hear the little voices inside both our heads shocked by this behavior, insisting we stop. If we had been caught, we could have been expelled from Cascade and sent to a lock-up, or worse, put on Bans with the entire school and forced to dig ditches for weeks and weeks, all privileges and respect rebuked. We would have been a disappointment to ourselves and to the school. We stopped suddenly. The guilt had triumphed. We nodded goodnight and returned to our separate tents.

After that night, we never shared another sexual interaction, even though we did spend an excessive amount of time with each other, taking long swims around the perimeters of lakes or washing our clothes in the stream as we sipped hot cocoa for an unusually long period of time. A week after we returned to Cascade, Mike and I were put on Bans, though the counselors had no proof of sexual activity. They'd sensed something between us and decided to extinguish it before it caught fire. I didn't speak to Mike again until the following February, the day before I left Cascade forever.

Three years after graduation, we met again and entered into a two-year relationship. But we never shared another night as passionate as the one inside that sleeping bag. Perhaps because our love could only thrive in a place wholly detached from the world's chaos. In that moment, when all that mattered was right there in front of us.

22.

LIKE FLOWERS NEED THE RAIN

In early September, five months before my graduation date, I found Lu sitting Indian-style in the middle of her garden picking petals off flowers she had planted that summer. She wouldn't speak or look at me. Her hair was knotted in a ratty ponytail, her face smeared with dirt, and she rocked back and forth, humming a song I didn't recognize. A collage of electric yellow, bright purple, and baby pink petals surrounded her. The earth seemed to burst with color. I knelt beside her, stroking her long, bony back, my fingers rising and dipping with each vertebrae as if they were mallets across a xylophone.

"Lu, what happened?" I asked.

She didn't flinch, didn't look at me, didn't acknowledge my words or my presence. I asked her again and again, but got nothing, just the unfamiliar song being hummed over and over. I lifted her into my lap. Her muddy, lifeless legs flopped over my arm, her body entirely dead weight against mine. For a moment, I thought she'd gone crazy.

"Please talk to me. Please tell me what's wrong. You're scaring me," I begged.

After what seemed like ages, she turned her head and stared up at me with those eyes capable of melting a glacier. Tears poured from them, ran down her cheeks and formed little puddles in her clavicles, but she spoke calmly.

"I confronted my parents about Austin, told them everything, but they said I was a liar, and they said I better not say anything like that again or I was going to go straight to Hell. They didn't believe me, and they think I'm nuts. They think I'm making it up. Who would make something like that up?"

"Nobody," was all I could think to say.

"What is the point in fighting if I have to return to him?"

And what could I say to that? What does one say about such an evident crime against human justice when its victim is curled into a ball of tears in your lap? I was unable to utter a single sentence of consolation, justify it, or solve it for her as I always had. She felt breakable in her frail body, her small, round head pressed against my chest. I remember thinking how strange it was to grieve for what was to come, to know your future held such inescapable pain.

We stayed in the garden until it was dark, and then we watched the stars blink and pop, shooting from one corner of space to the other. The sky was bright with color, and the commotion distracted Lu for a bit. Her head rested on my stomach, my head on a moldy log. I knew we should have been inside getting ready for dinner. I knew I was breaking a rule, and it was wrong of me, but I didn't want to disturb the brief moment of peace she'd found.

"What can I do?" I finally said.

"Just tell me you're scared, too."

"Of course I'm scared."

"What are you scared of?"

"That," I said, gesturing toward the cold universe above us.

"Yeah, me too," she said.

We lay with the earth for a while, and she braided wreaths from long stalks of grass. When she finished she placed one on my head and one on hers and said they were our crowns, and then without warning, she began laughing and pulled herself up, extending her arm for me to take.

"Why are you laughing?" I asked her.

"What else am I supposed to do?" she said, linking my arm.

Lu's favorite spot

And as we walked back to the dorms with circular green braids wrapping our heads, she hummed, *I need you. Like the flowers need the rain you know I need you.*

Thinking back, I guess she was right to laugh. Her laughter was a kind of acceptance of the iniquitous world, a realization that one must go on despite the horrors. She was allowing herself to rise from the trauma of what was to come. We both knew fighting would only worsen her situation. Though Cascade believed Lu, the school was helpless because if it chose to fight her parents, if it involved child services, Lu would have been taken to a home until she was eighteen. Her counselor insisted her parents discuss the molestation with Austin. When they did he was shocked, furious to be accused of such a thing. The parents believed him, and it was put to bed.

*   *   *

Because I had become an exemplary Cascade student, I now did Move-ins. New students arrived about four times a week, usually coming from wilderness programs, lock-ups, or previous boarding schools. Some were sad and impassive, others quiet and terse and insistent on ignoring my efforts to befriend them. Most, however, were amenable. The day I moved Jillian Pitts into Cascade was unusually busy in the admissions office because two other girls were arriving as well. Jillian stood next to her escort with both hands on her hips and a cocked head, already bored with the scene. She came from West Hartford, Connecticut. She wore a tight orange fleece vest over a tighter white T-shirt and torn jeans. Her lips were chapped and stained with cheese doodle powder. She had thin, oily, pin-straight hair that hung loosely around her face and beady green eyes, and she interrogated me on every aspect of the school as we walked toward the Welcome Center.

"Hey, little lady. What's your name?" Marla greeted us with her big grin.

"Jillian Pitts," she said.

"Want a sucker?" Marla asked.

"Love one," said Jillian, with abundant sarcasm.

"Do you know the routine, Jillian?" Marla asked.

"No, she doesn't," I said. "Janice was busy dealing with another Move-in when she arrived so the secretary just told me to bring her over here."

Marla went on to explain the strip-search process and Jillian's nonchalance vanished. Her cheeks flushed, and she took a step back and stared down at the caramel sucker still in the wrapper. This was my cue to put my arm around her and reassure her that everything would be fine, just as Rona had done with me nearly two years ago.

I did. I promised Jillian it would be quick and painless. I mentioned nothing of the humiliation she would feel. I mentioned nothing of the shame.

I no longer remembered it.

She shrugged her shoulders, unwrapped the sucker, popped it into her mouth, and followed Marla into the bathroom without a word.

After a few minutes, the front door opened and Lara came in trailed by a thin girl with bleach-blond hair. My eyes locked on her face, the bulging brown eyes and long lashes, the high cheekbones and large, pimply forehead. I knew this girl—although without the Spence uniform it took me a moment to place her—Elizabeth Marks. She'd been in the grade below me at Spence.

"Elizabeth," I said, smiling.

"Hey," she said. "I didn't know you were here. I didn't know what happened to you."

"They kicked me out."

"I know that, everyone knew that, but then you vanished from New York. I never saw you out or anything," she said.

"Well, here I am," I said, a bit shaken. "Now you know."

Marla walked out of the bathroom with her arm around Jillian, whose face was red and puffy. Elizabeth turned toward me for consolation, hoping I could save her from whatever this weepy-eyed girl had just endured.

"It's the strip-search," I said. "Don't worry, it will be quick and painless."

"Fuck that," she said, rearing back.

"Okay," Marla said. "Alison, can you please take Jillian with you to tour the campus and then come back later for her belongings and," she turned toward Elizabeth, "dear, you come with Marla."

Marla signaled with her hands as if she were telling one lane of traffic to go and the other to stay. Lara put her arm around Elizabeth, who shrugged her off with the jerk of a shoulder and glared at me with widening eyes.

"You're moving that girl in?" she said, outraged.

"Yeah," I said. "I know it seems strange to you now, it did to me too, but you'll get it soon enough. You will grow to love it here, Lizzie."

"Fuck that," she said again. "And you of all people. Unbelievable."

Seeing me there in my clean, navy sweater, precisely ironed corduroys, and classic loafer-type shoes, my hair neatly pulled into a ponytail, no makeup, no earrings, and above all, advocating Cascade, astounded Elizabeth Marks. She stood there unable to peel her eyes off me, as if staring at me long enough would remind me who I really was. I grinned sheepishly, linked my arm through Jillian's, and left her in the Welcome Center with Lara.

Seeing her reminded me of the life I had lost, and for a brief moment I felt nostalgic and slightly embarrassed about my submission to Cascade. I began to wonder who I had become, but then I realized that those thoughts were my negative voices talking, my "I," as Cascade had taught me in the I and Me Workshop. I needed to listen to the good, the true, the "Me." I was so steadfastly dedicated to the program by that point that within seconds I was able to block out any questionable thoughts about my new self and the mission I was on. I remembered the "bigger purpose." I remembered Mira and her belief in me, and I remembered our fight for a sane, caring, and enlightened world. And I turned my attention to poor Jillian.

Elizabeth Marks rebelled for a few months but eventually became a devoted Cascade student just like the rest of us. We never talked much, never discussed our interaction that day, but ended up as roommates in the East Village years after we both graduated.

23

I AM A TRUSTING AND HONEST WOMAN

My final Workshop at Cascade was the Symposium. It was the longest of the eight Workshops, four days, three nights, and then presto—a new you. The Symposium was advertised as an apocalyptic event, something that would alter our lives forever, transform the way we viewed the world and ourselves. Even after students had been through it and knew every infinitesimal detail that went on over those four days, they still spoke of it as though it was some sacred rite of passage.

"I learned more in those four days than any other four days in my entire life."

"I wish I could do it every five years."

"I would have definitely killed myself if wasn't for the Symposium."

Of course I fell victim to the lure of the unknown, creeping about days before my Family was to begin the workshop, trying to catch a tidbit here or there, desperate to know what I was about to experience.

The Symposium:

We sit in the usual horseshoe shape, the one every workshop begins with, but this time we hold notebooks in our laps, and this time, we know, is our last. My palms are moist and clammy and smell like metal from the spirals on the outside of the notebook. An easel stands at the front of the semicircle with an oversized pad of paper. On the paper is a chart drawn in black marker that looks like a child's maze. It is called an Emotion Chart. The words *disgusting, victim, sell-out,* and *inflexible* are written on a path that seems to either lead into or out of the maze, Inside the maze are more words: *ugly, isolation, hatred, lies, blame, denial, failure, cruel, weak, coward.* And if you go deeper into the maze, there are more words: *shame, hopeless, fear* and *anger.* Then in the center, a circle of good words float freely like bubbles: *acceptance, power, generosity, love, honesty, kindness, beauty, courage, responsibility, trust.* They tell us to find our home. Do we live in hopeless, fear, anger, or shame? And then they ask us what two words will get us out of there, what two words from the free floating circle will allow us to escape this maze of misery?

I stare blankly at the chart. The words slip into my head and float around in my mind like strings of seaweed on the ocean surface. They leave no impression. All the words seem to have related to me at some point in my life, but none seem specific enough. I think about fear and anger. I can't decide between the two. I can't remember a time when I wasn't angry at my mother or my father or myself. Finally I choose anger.

I write neatly in red ink on the first page of my notebook— A-N-G-E-R. I am careful with each letter, careful to make the word neat. I trace over the letters until they are thick and wet and disintegrating the paper, then I leave the word sitting alone on the big white space. I stare again at the chart. Anger leads to stubborn, cruel, hurtful, resentment, righteous, controlling, and inflexible. They tell us this is probably where we live when we are at our worst. I picture myself like Alice in Wonderland after she bit from the mushroom. I am tiny and sitting inside the maze with the words. The space is like a room in a dollhouse and these words circle the walls. *Hopeless, victim, cruel, coward, hurtful, desperation, blame, denial, guilt.* Not for a second do they let me forget where I live. I want to scream or cry, but instead I laugh cynically, and the counselors flash me looks of disapproval.

Now I must think of words that will save me from this room.

This part is easy. I choose trusting and powerful. They will get me out of my room, out of my life. They will complete me. One at a time, we stand in front of the room to announce our findings. The counselors nod or question, allow us to sit or keep us for further interrogation.

"We don't think you've chosen the right words, Alison," a tall man says.

"We agree with trusting but not powerful," a woman concurs.

I know I have chosen the right words. I know I desperately need powerful. I know that I am scared to speak, scared of being wrong all the time. I know I feel weak every day, but I also know the counselors won't let me sit down until I appease them.

"We think honest would suit you better. If you are honest with yourself, you will become powerful," Katherine says softly.

"If I am trusting, I will be honest. I need powerful," I say.

"No Alison, your statement should be 'I am a trusting and honest woman.'"

I still don't agree, but I believe they know best, and I also know I must play by their rules in order to graduate in two months. I nod and announce my statement to the room. They say this is my life-long contract with myself, and if I live by it, I will never believe my negative statements, my "I." I nod and sit down.

The next exercise is a costume party. We must come as a character that represents our worst qualities, the people we become when we do not have the words that we just picked. The counselors have chosen our characters, and we must guess who each of us is. Katherine runs into the middle of the room with a bunny costume draped over her arm. She repeats something over and over again; I can't understand what she says. She has a timer around her neck, keeps sniffing and running and sniffing. "I'm late, I'm late for a very important drink. I'm late, I'm late for a very important drug. Don't look back. Go, go, go." I now understand what she is saying, what she keeps saying. She runs up to people and asks them how they are, tries to take care of them and help them, but she refuses to speak about herself, and if someone asks her anything, she turns away. The timer only allows her three seconds a day for herself. Beep, beep, beep, go, go, go, forget, forget, forget, numb, numb, numb. I know right away she is acting out my character, and I raise my hand. She hands me a costume, a furry white rabbit suit with pink ears. I am the white rabbit in Alice in Wonderland. My hands wring the ears. I am told to change into it. My hands twist the ears harder.

An hour later everyone is dressed in costume and running, like lunatics, around the room. Sam is Peter Pan. He dresses in a green leotard, a green felt hat, and green elf shoes, and instead of flying he skates on a green skateboard while snorting cocaine from his hand. He can only speak in baby-talk. Marilyn is the Bag Lady of Broadway; she dresses in rags and lives as though she is perpetually on stage. She dances and sings around the room, but occasionally she falls onto the ground and sobs loudly. Mike is Raggedy Andy, and he flops around silently, cuddling up to people and speaking to them in his whisper-like voice. Lara is Cinderella. She scrubs floors and waits on people because she doesn't believe she is worthy of anything else. Jake is Superman and Clark Kent, a boy capable of extraordinary things but always resorting to the weaker person inside himself. And Crissy is a slutty Tinkerbell; she wears a minuscule tutu and cropped tank top and carries a magical wand that she uses to sprinkle fairy dust on people and make them fall in love with her. One girl is the Evil Stepmother, she walks around the room jabbing people: Your mother left you on the doorstep. Nobody wants you! she says to Lara. You faggot. You loved it when daddy touched you! she says to a boy who was molested. And as usual a song blasts from the stereo, the music lifting and amplifying our emotions.

People cry and scream, hysteria enters the room. I feel insane. I feel hot and sweat a lot and think how stupid. I am terrified that I feel this way because I know it is this Workshop that is supposed to save me. I try to throw myself back into the midst of it, to force my belief through action: I run up to people and try to take care of them. I want to please them. I want to help them. And I refuse to talk about myself when they ask me questions. I pretend to drink and snort cocaine, and I stop every few minutes and give myself three seconds to reflect on my life. The evil stepmother tells me I am killing my father. She says I am a failure to him and everyone else. She says I caused my mother so much pain that she became an alcoholic. Mike comes up to me and tries to cuddle. Sam skates by giving me a thumbs-up sign and nodding. I feel like I am losing my mind and wonder if anyone else feels that way, but I keep running and running. And the music continues.

"Stop, and take your seat!" a counselor yells.

"This is your worst. This is you living out your negative statement. If you live by your contract, you won't be this person. Let's fight away this person. Let's really destroy it!"

Pillows are brought out. Socks are brought out, some with the faint outline of old blood stains. Chairs are stacked in the corners. The room smells like spit and insanity, and I can hear the rise and fall of chests, the lingering whimpers and the old tape player clicking.

"And *go!*" a counselor screams.

We no longer need an explanation. When pillows are brought out, we know what to do. We are the proverbial dogs to Pavlov's bell. Within seconds the socks are on our hands and we are pounding, pounding up and down. I feel my entire body pressing against the wall of my face. Pressure builds and builds, and the blood vessels around my eyes begin to pop.

"Go away! Leave me alone!" I scream. "I hate you! I hate you!"

I begin to rip at the neck of my bunny suit. Music blasts louder now.

Sam is lying close to me in a fetal position, and his hands are balled into fists that slowly pound his forehead and I can hear him whispering something over and over again. I know he's not fighting. I reach my hands out and grab his fists to stop him from hurting himself, and he opens his eyes and we stare at each other, and for a minute it feels like we can see inside each other, like we know each other in a way uncluttered by actually knowing each other in the context of real life. He inches closer to me, and his head burrows into my stomach, and his body begins to vibrate. I rub his back as it convulses up and down. I feel my T-shirt soak with tears. The louder he cries, the tighter I hug him because I think, the closer he is to me, the farther away he is from himself.

"Okay, stop!" Katherine finally yells. "Now one at a time please stand in front of the room and announce who you really are, your contract with yourself. Please, only come when you are ready and when you really mean it."

Many people try and are shot down. I go, and Katherine says:

"Alison, would you rather walk into a room, stay for awhile, and when you leave have everyone say, Wow, who was that girl? Or walk into a room and when you leave have one person say, Thank you?"

"One person say thank you," I say, thinking of Lu.

"And Alison, do you really, truly understand why you must allow yourself to feel things, even pain? Why you can't just run through life and never allow anything in?"

"Yes," I say.

They believe me and tell me to state my contract. "I am a trusting and honest woman," I say, but I don't really believe myself, and I know that, but I don't care anymore. I'm tired and hungry, and I want it to be over. I am praised, patted on the back. I am a trusting and honest woman.

After everyone says their statements in front of the room, we are told that it is time for the Rocking Exercise. We are told to line up in pairs facing each other and grasp our hands together in a crisscross formation so we form a hammock of hands. One by one we lie down in this human cradle and rock each other. Personalized songs play on the stereo while we are being rocked. Some people cry, and we stare at their faces that seem to stretch and fold like silly putty into themselves. I can't cry, not with all those faces staring down at me, so I close my eyes and listen to the words of the song. The words talk about the love in our hearts and how they will make the world a better place.

They sound nice, but they make me angry because they are a lie. Love in my heart cannot help Lu, and it cannot heal the world. As my song ends I open my eyes, and everyone is smiling and crying happy tears. I tell myself to stop doubting this Workshop. I remind myself that this is my last chance. I struggle to believe the words in the song. I believe the words in the song.

After we are rocked, we do the Flying Exercise. The counselors say we can now fly. Our contracts have made us so powerful we can fly. Everyone extends their arms, one by one we lift people into the air, our hands flat against their backs, their legs, their heads. We walk them around the room. When I fly I feel free. I believe I will succeed if I keep my contract. I am a trusting and honest woman.

The next three days blur for me. Yelling, crying, screaming; some I really feel, some I fake. On the last day, we put on pretty dresses and comb each other's hair and prepare to return to the school. We walk holding hands like a long chain of construction paper cutouts, and as we move up the hill, over the bridge, and down the gravel path, I hear music coming from the House.

For a brief moment, I believe we are all fixed. Our cheeks are rosy, and our eyes are lucid with post-crying clarity. We look as if we have just glimpsed our amazing futures, and we are now ready to embark on the journey. Sam, Jake, and Mike wear pastel button-down shirts and khakis and hold hands and smile big, effortless grins like brothers ready for a family photograph. Sam's normally brown hair is sun-washed from his summer days spent running shirtless through the woods, and it blows in thick, perfect blond curls as he walks. Lara wears a green and white flowered sundress and squeezes my hand tightly as we near the house, as if to indicate that we've made it. She glows brightly that day, even without her fake eyelashes and bright lipstick and big, sparkling earrings. Crissy leads the long line, her blond-red hair wiping her face and catching in the strawberry chapstick she always wears. Marilyn is last, and when I turn to look at her, she is already smiling and waving at the people she can see waiting for us back at the House. We are amazing that day. We are capable of changing the world.

Cat Stevens blasts from the speakers: If you want to sing out, sing out! he tells us.

When we reach the House, I can see the entire school waiting for us in a semicircle. I spot Lu in a white sundress with her hair combed down and parted neatly on the side, her usual red headband holding it back. She is smiling, and I want to cry when I see her, but I know I can't, so I smile and grab her hands pulling her to me. Together we skip in circles.

Everyone dances and sings along to the music, and Lu's hair flaps in the wind, and I feel happy, truly happy and free and full of love. For a moment, all my worries and fears dissipate, and I believe Cat Stevens when he says that I can do whatever I want and be whomever I want. Jake and Sam spin each other in circles, and Marilyn dances with her Little Sisters, twirling them under her arm one after the other. She looks at me and sings—*If you want to sing out, sing out,* and I sing back—*if you want to be free, be free.* And she dances over to Lu and I with her Little Sisters trailing behind, and we all join hands and skip in a circle.

I look for Mike, but I can't find him. He is not dancing. He is on the sofa with Katherine, and she holds him, and he cries hysterically. He is not happy or free. He is not smiling anymore. He doesn't believe his contract. This was his last chance, and he knows it. Lu tells me I look beautiful, and I tell her I feel beautiful, but I don't anymore. I can't keep my eyes off Mike. And suddenly I remember that Cat Stevens lost his mind and was unable to function in the great world he is singing about and that he escaped it by joining the Muslim faith, retreating from reality and changing his name.

Later in the evening, I am alone in the dorm, and I stare in the

Cascade students assembled to welcome back a Workshop

mirror. It is one month and three days before I leave Cascade forever. I see the same pale, transparent skin, the rosy cheeks, the blue
eyes framed with broken blood vessels and thin, arched eyebrows.
I look no different, but I try to convince myself I feel different; I am
a trusting and honest woman—right?

PART III

SUCH DESPERATE, DIRTY PEOPLE

One ought to sink to the bottom of the sea, probably, and live alone with one's words.

—VIRGINIA WOOLF

24

THE WHITE RABBIT RETURNS

Freedom, when given suddenly, can be a terrifying gift. It rushes at you like water into a gulf, and if you don't swim with it, you lose yourself. Thrown back into the world after two years of isolation, I quickly saw what a cold and unforgiving place it could be. My formerly hardened self, once inured to disillusionment, had been softened, and I was unequipped to handle reality. For my first few weeks home I existed in a state of terror as the world in its chaotic fury whirred around me. I didn't leave my apartment very often, and when I did, I was scared of everything: crossing streets, speaking to people, pigeons in large groups, men who stared too long, women who spoke too loudly, any figure of authority.

Sitting alone in my bedroom night after night, I wanted nothing more than to return to Cascade. It was a terribly lonely white-walled room with hardly anything in it: a few china dolls, old Spence yearbooks, my blue-and-white bed with its monogrammed baby pillows, an empty desk with a framed picture of my family, and a dresser with my silver baby cups on top.

My father was eighty-four when I moved back home in 1996, and Cascade still wanted me to change him. The counselors had deemed him a detriment to my personal growth, saying he was offensively critical, degrading, and a complete narcissist. They said his incessant belittling was damaging to me and would never allow me to flourish. I believed them.

I believed that if I didn't confront him, if I didn't refuse to accept the only man he was capable of being, I would be destroyed.

I confronted him over lunch one afternoon a week after I'd returned home, spewed out some therapeutic garble that he didn't understand. He looked back at me across the bread basket and the blue glasses filled with ice water with an expression of genuine confusion and hurt.

"Well, this is news," he said. "I didn't know a daughter could break up with her father."

"I'm not breaking up with you, Dad. I'm simply saying that if you're unwilling to change, a relationship will be impossible."

"Change what, Alison?" he said, sincerely. "The jokes we've been laughing about together since you were three? I'm not allowed to tell them anymore?"

"You just don't get it. You never have, Daddy."

"I suppose I haven't," he said, staring at the ice cubes.

A few weeks after I arrived home, I went out for dinner with my old group of Spence friends at one of our uptown haunts. They laughed and gossiped and argued just as they had three years ago, though now I had nothing to contribute. I didn't get the inside jokes but laughed anyway. Once in awhile someone would chime in with a question for me, but on the whole they didn't seem to care that I was home.

"So, Alison, weren't you at like a totally crazy nuthouse?" Niccola said, fiddling with the Tiffany heart hanging from her neck.

"Sort of," I said. "It was pretty strange."

"Well, did it fix you?" Samantha said.

"Or should we keep you away from delis that sell coffee beans?" Niccola blurted.

They all burst into raucous laughter. I forced out a quiet chuckle, but my real laugh was loud, and they knew I was faking. I couldn't fool these girls; they'd known me since I was five years old. And as the meal continued, they began to look at me strangely, examining my clothes, my lack of jewelry, my new polite, introverted self. They knew I wasn't the same

Alison Weaver who had left them, and they were right. Sitting there that afternoon among tables of young girls in various private school uniforms, among high-pitched giddy laughter, Diet Cokes and Shirley Temples being slurped, and voices bragging about all the little successes they'd totted up over the last year—I realized I had no idea who I was.

Marilyn and I had returned to New York together. We'd left Cascade in the same medicinal white van we'd arrived in with the same Christian rock station blasting from the speakers, and we'd vowed to stay true to the school's ideals. For the first few weeks, we wholeheartedly believed that if we did as Cascade had told us, we'd be saved from the temptations of the immoral world. I picked a new Value to live by each day from a card of values we'd been given in our Heroes Workshop: *integrity, honesty, truth, acceptance, strength.* I did everything I was supposed to. I didn't curse or yell at my mother. I used "feeling" words to express myself. I ended my relationship with my belittling, sexist father. I stared into my bathroom mirror every morning and stated my Symposium contract. I had it written on a piece of loose-leaf paper and taped in the center of the mirror. In the morning before brushing my teeth I read: *I am a trusting and honest woman. I am a trusting and honest woman. I am a trusting and honest woman.* Sometimes I even said it to my reflection in store windows on Madison Avenue or bus windows or the rearview mirrors of taxicabs because they'd told us if we stated our contract every time we began to doubt ourselves, every time we felt alone, or every time we felt like taking a drink, we would be saved. We would be keeping our Symposium characters at bay. If I believed my contract, the White Rabbit couldn't get me. But they were wrong. It didn't work, and eventually I gave up trying.

It was surprisingly easy to break the two-and-a-half-year-long sobriety. I walked into a bar on Fourteenth Street, ordered a glass of the house white, and down it went, smoothly. The small amount of guilt I felt was mollified by the second drink, and by the third I couldn't remember why I'd stayed sober for so long. In the future, I would find it just as effortless to smoke pot, drop ecstasy, snort speed, cocaine, and crystal meth. My parents had hoped I would return from Cascade and be back on the track they had laid out for me years ago. They'd hoped I'd be ready to pack up for some posh Ivy League school, pick up with some wealthy accountant or lawyer with straight, pebble-white teeth who wore checkered Brooks Brothers shirts and navy blazers with gold anchor buttons, who used expressions like "summered" and "nightcap" and "up for a game."

But that life made me sick.

DECLINE

I went to college because that seemed to be the thing to do. I attended Pitzer, one of the Claremont colleges half an hour east of Los Angeles. California was warm; it had that alluring sixties generation mystique and seemed to be the spiritual center of self-discovery and change. It was also where Mira, who had only remained sober for one month after leaving Cascade, attended college.

The nineties rave scene was alive and flourishing in California, and it sucked me right in. My daily attire consisted of oversized, pink fuzzy overalls or neon-colored bell bottoms and tight baby T-shirts, pastel barrettes shaped like various cartoon characters, and a purple pacifier that hung from my neck. In the apartment I shared with Mira on Bonito Avenue, we finger painted the walls, hung hundreds of colorful beads from the doorways, and posed naked in Adidas platform sneakers on our balcony. We drank from baby bottles, wore little boys' Transformer underwear, and brushed our teeth with electric Teenage Mutant Ninja Turtle toothbrushes.

On weekends we took ecstasy and attended parties called Techno Babies or Moon Tribe and sat in massage circles as music thumped from enormous speakers: trance, jungle, hardcore techno. Dropping a pill of E was like putting on a coat of heaven. Life became absolute bliss for those four or five hours. I loved myself, every little thing about myself: the lines under my uneven blue eyes, the sound of my nasally voice, my rhythmless, sloppy dancing. And I found unquestionable beauty in everyone and everything I interacted with. I was confident and open, and I remember

At a rave in California with friends

one night I stared at myself in the bathroom mirror of some ratty venue in Los Angeles and thought—*I finally believe my Symposium contract, I am a trusting and honest woman. I really, truly am.* I felt okay in my own skin, better than okay. I was happy.

Some nights we drove for hours into the middle of the San Diego desert and danced for two days straight until the cops arrived in helicopters and escorted hundreds of tripping teenagers in candy-colored clothing to their cars. California was a new beginning for me. Nobody knew who I was or where I had come from. Nobody expected anything of me. I could finally be anonymous.

Mira and I spoke so highly of our new life that Marilyn, who was failing out of Yale, soon dropped out and moved in with us. Within weeks she was addicted to speed, and often at night she'd lock herself up in the bedroom, collaging her walls with rave fliers and various photographs. When I went to the bathroom in the middle of the night, I could sometimes hear her talking to herself. I could hear the rip of scotch tape, her feet moving across the tacky shag carpet and the photograph being pasted up on the wall. In the mornings, as Mira and I were getting ready for classes, she'd be retouching her makeup from the night before, sniffing wildly to prevent her nose from running.

When we got home at the end of the school day, she'd be lying across our secondhand thrift shop sofa in a full-piece white cotton pajama outfit with plastic feet, smoking a Camel Light and sipping a fruity cocktail out of a Kermit the Frog cup while the television played some midday talk show.

"How was class, girls?" she'd ask us if she wasn't too engrossed in the show.

"Swell," I'd say and plop down next to her.

"Where are you hiding the drugs?" Mira would ask.

"Get your own," Marilyn would say. "I paid good money for these bags."

She'd sit up and pull three or four small transparent green bags of speed from her bra, dropping them on the table.

"These cost me six shifts at that awful Italian restaurant—six shifts!"

"Well, if you weren't high all day, you wouldn't have to work so much," I'd say.

"But I wouldn't be high all day," she'd say, pulling a long salmon-colored string of gum from her mouth and smiling.

Mira and I had limits and agreed that if we stayed within them, we were simply recreational users, not addicts. *Weekends only* was our rule. But soon weekends trickled into Monday and Tuesday or began on Thursday, and within a few weeks we were calling the dealer whenever the hell we so desired.

Most evenings cheap box wine would be brought out from the fridge and poured into colorful plastic cups. Some neighbors would trickle into the apartment. A blond boy I was dating named Donnie would sometimes drive up from L.A. with his twin brother Marco. Lines would be snorted off the mirrored coffee table. We'd watch a few movies, set up a photo shoot in the parking lot, maybe pop on a new album by the Lords of Acid or the Chemical Brothers, and dance. Marilyn would go into the bathroom, fill the tub, and splash her foot around in the water while she sat on the toilet seat retouching some cheek glitter. She wanted us to think she was bathing, but we knew better. Marilyn didn't like to bathe. She worked so meticulously to perfect her makeup it seemed a waste to wash it away after only one day.

Soon cocaine and ecstasy were no longer enough, and I was turning to speed and crystal meth. On speed, I was paranoid and mean. My eyes bulged from their sockets, my gums were red and numb, my jaw moved unnaturally, and my nose adopted the habit of bleeding periodically throughout the night.

By the end of the school year I was visibly unwell. My weight had dropped twenty pounds; I had permanent ash-colored half-moons that hung under my bloodshot eyes; and I had trouble concentrating. I could barely sit through a lecture at school without nibbling my nails to the quick. The initial excitement and freedom of the unrestricted party lifestyle had grown old and boring. I was tired of going out. Sometimes I was able to force myself to have fun, but often I felt myself shrinking to a small dark speck against all the thumping noise and glittering color of the party. My cocaine mouth tasted like dirty underarm sweat. My throat hurt. My glands were swollen. And I found something unsettling about watching all those people laughing and bouncing about wildly and petting each other. If I allowed myself to think about it for too long, I became very lonely. It was like watching something that once appeared heavenly and promised happiness turn dark and morbid and then implode.

Two years later, the three of us decided to transfer to the New School University and return to New York City. That September I was hired to teach reading and writing to first-graders at a small East Village school called Saint Thompsons. I thought I'd be able to clean up my act and have a real life, but it didn't quite happen like that.

It was around this time that Mira and I began to have problems with our friendship. In Mira's version of the story I became a reckless drug addict who she could no longer associate with; but Mira's version of most people's lives is slightly skewed.

The truth is we ruined ourselves together, and then we woke up one morning and couldn't stand the sight of each other. I stuffed myself with vacuity, and developed a wonderful capacity for passive detachment.

At a party, high on K

Mira turned aloof and vengeful. Her personality, which had once been so kind and comforting, so reachable, turned cruel and distant. When we looked into each other's eyes, we both saw strong, beautiful women in ravaged bodies nearly destroyed by addiction and depression, so we ran the other way.

In August, I flew to Redding to greet Lu at the airport. She was graduating from Cascade, moving back to Florida where her parents and brother waited. Standing in the gift shop flipping through a diving magazine, she still looked like a little tomboy in her baggy jeans, red sweatshirt, and untied Converse sneakers. I snuck up behind her and placed my hands across her eyes.

"Al," she yelled. "You came!"

"Of course, I came, silly," I said.

"Look at these poor dolphins," she said, shoving the magazine in my face. "They're trapping them and bringing them to private hotel zoos. It's so wrong. Look at their faces. They're so sad to be leaving their homes."

"Welcome to the world," I said.

"I have to buy this magazine. I'm going to write these awful people," she said, looking me up and down. "You look scary, Al."

"Don't be stupid," I said. "I'm just not in Cascade dress code."

"Are you doing drugs?" she asked.

"Of course not," I said.

But I could tell she didn't believe me. She wanted to, but her keen intuition wouldn't let her. We went upstairs to the Chinese restaurant and shared a bowl of white rice with soy sauce. After we finished eating, we curled up on the sofa booth, and I held her just like we used to in the corner of the Cascade House. It felt good to connect like that with another human being again, and for a second I missed Cascade.

"Do you think about it anymore?" she asked, her big, wet eyes staring into me.

"Every day, Lu," I said. "Every day."

When her flight was called, I walked her to the gate, wheeling her heavy red suitcase behind me, and she clung to my waist with her string bean arms.

"See you, Al," she said. "Love you so much."

"Bye, Lu. Love you, too. We'll see each other soon. Stay strong."

"Yeah," she said.

She took her suitcase from my hand and tottered down the gateway, her long, knotted red ponytail swinging across her back, her frayed jean hems trampled and muddy under her heels. That was the last time I ever saw her.

We spoke on the phone a few times that year, and then we drifted. When I tried to reconnect a few years ago, leaving messages on her machine, she never called back. Last I heard she was living in Oregon, maybe saving dolphins or endangered species. But whatever she's doing, I am sure the world is a better place because of it.

XERS

I rented a six hundred square foot studio in a prewar building on Second Avenue between Eleventh and Twelfth Street. It was on the ninth floor and had a dark and gloomy view of three acrylic-painted brick walls. The space was haphazard but quaint; it had dark amber floorboards, a shiny green marble bathroom, a walk-in closet, and a Greek-style arched entryway. I decorated it in a kitschy, modern style: two large, clear blow-up chairs, a red and black airplane seat sofa, brightly colored transparent plastic stools, two mirrored tables, a giant blue plastic jacklight, an Elvis lamp. A chest of drawers separated the living room from the makeshift bedroom that contained nothing but a futon mattress with leopard print sheets and a few large faux fur pillows.

I wasn't very good at living alone. Many roommates came and left. The one permanent tenant was Astro Earl, a thirty-five-year-old man who worked as a host for what was left of the downtown club scene. He had shoulder-length brown dreads that he dyed various shades during our time together and a tan, pocked complexion. His cheekbones were high and his bark-colored eyes popped out at you every time he blinked like a thick yellow explosion of comets. Each night he would pluck his eyebrows with precision until they were arched into a fine line above his eyes, gloss his lips, and dust a thick, black brush covered in loose beige powder across his oily face, even if we were staying in.

Some nights he'd dress as a female devil with a long pleather gown, flaming orange sleeves, and a red bodysuit; other nights he'd dress as a

Astro Earl and I in one of our photo shoots

flapper with a black ostrich feather jacket and Indian bindis tracing the arch of his eyebrows. Then there were the nights he covered his entire body in nothing but a sheer body suit and silver glitter. On his feet he wore the highest black stilettos I'd ever seen.

Astro put the letter L in front of all words beginning with A. Alien was *L'alien*, and Africa was *L'africa*, and apple was *L'apple*. He also added Tress onto the end of certain words that he felt sounded better with it. Sweater was sweatress, and water was watress.

When he wasn't working, Astro wore T-shirts that said ANARCHY across the front and camouflage cargo pants. He carried his clothes around in garbage bags, and his most prized possessions were his plastic art cases that stored hundreds of lipsticks, eye shadows, and powders. There was a hardness about him, a detachment from reality that both scared and enticed me. His personality was domineering and intimidating, yet if he felt safe, his features softened and he'd relax into himself, revealing a gentleness and sensitivity few knew.

There were about seven other people who frequented my East Village studio over the years. Various village stragglers I'd picked up along the way and used to make myself feel like a good person. I'd give them Cascade pep talks, elevate their morale, offer my apartment as a place of refuge until they got on their feet. Then two months later they'd be addicted to some drug, and I'd be bitching about them to anyone who would listen.

Mikey moved in only four months after Astro got there. He came from a farm in Oklahoma. I met his father once. He was missing four front teeth, and the few he did have were brown and spongy. I'd gone to a family wedding as Mikey's heterosexual beard and his father, drunk and thinking he was doing his son a favor, rented us a room at a hotel on the freeway with a flashing neon sign saying Honeymooner's Haunt. Our room had a bed with pink silk sheets, a mirrored ceiling, a disco ball, and a basket filled with complimentary flavored condoms.

Mikey was a fashion stylist and model. He made decent money but spent it all on dope. He was tall with large feet, long, thin spider-leg fingers, and spiky black hair threaded with lights of blue. On the pages of the glossy magazines he made it into, he looked sweet and helpless, and for a while I believed in that part of him, though in person something dark and scheming began to seep through. Every month he looked dirtier and sicker, a sneer appearing in his expression, hollow rings deepening below his eyes. Still, in the magazine spreads he appeared innocent and beautiful; and still, I tried to convince myself that the Mikey in the magazines was real.

Then there was Nellie, a beautiful, African-American drag queen who'd been infected with AIDS at seventeen. Her hair swung across her back and around her face like black straw, crunchy and inflexible, and her eyes were watery and brown, like a sick dog's shit. She was 6'2" and walked with an elegant strut around the apartment in her pastel-colored teddies and hairless, pelican legs.

Nellie was in love with Baby Joel. He hung around the apartment sometimes too but mostly he slept in the park under the yellow slide. Young and handsome with soft, pale skin and sexy, gloating eyes that, when bestowed upon you, made you feel your drug use was somehow inadequate to his. Which in fact it was. Nobody could keep up with Baby Joel. Jenny, his girlfriend, was tall and lanky with stringy, blond hair, a mild lisp and a dim-wittedness you wanted to hate her for but somehow couldn't.

Elizabeth Marks ended up there, too. One month after graduating Cascade she fell back into drugs, crystal meth specifically, and often we stayed up together for days at a time. One summer she got a letter from her father, offering to pay her five thousand dollars to spend the summer "where she belonged in East Hampton with her family." Of course she tore it up, but I remember thinking how that was probably the saddest, most disgusting gesture a father had ever used to attempt to reestablish a relationship with his estranged daughter.

Most weeks there were at least three people living with us in that tiny apartment. On some level, I believe I was attempting to create my own center of experimentation and creativity. I wanted what the beats and the pop artists had, a community of people who believed and fought for the same ideals, who understood each other and facilitated change and a journey toward something new and bigger. I wanted a Factory or a Studio 54 or a Max's Kansas City.

I had left New York City for Berkshire during the height of the glamorous and zany nineties club scene—a scene that claimed to be founded on the ideals of peace, acceptance, and unity. The few times I entered those clubs at fourteen and fifteen, I was whisked immediately away to a utopian dance floor with flashing lights, thumping music, and glowing, shiny faces moving like frenzied fish around me. My entire body would buzz with self-acceptance and love as I wiggled around in the crowd thinking that was happiness, believing that somewhere between the vapid conversations and drugged out hazes and smoke-filled rooms, true joy waited for me.

When I returned to New York in 1996, black-and-white posters of a drug dealer named Angel Melendez were taped all over Greenwich Village. I recognized him, had probably purchased drugs from him years ago at Limelight. Then his dismembered and headless body washed up somewhere along the Hudson River, and the murder came out, and

Astro Earl, ready for a night out

nothing seemed the same anymore. That period in the late nineties felt hopeless and dead and stood for nothing. The Village seemed filled with various groups of beatnik wannabes stumbling from club to lounge to after-hours party, searching for that feeling that would never exist, could never exist again. And I was right there along with them. It felt like we were a lost generation who had eagerly awaited our turn at greatness, who had served our time and were now ready to embark on fantastic endeavors, to help make whatever was supposed to happen next happen, but somehow that had gone wrong. No one cared. Everyone was tired and jaded, and it seemed like all hope and energy had been sucked from the world.

Many of the club kids and drag queens of Limelight fame, figures who'd once seemed magical and amazing to me, had turned out to be murderers or thieves. Everything was built on illusion, and morality had become ambiguous. The community I had so longed to be a part of as a young teenager was dead, so I increased my intake of drugs until I was able to convince myself that this was it, that we and whatever we were doing was the next big thing.

Our little family was like a group of shameless, negligent children. We shared an indiscriminate anger at society lacking in any logical calculation. It was just an apathetic decision to hate, because we had been hurt. Running from our oppressive pasts toward a future of something more liberating and less full of constraints and anxieties was the one essential common ground we all had. For some it was running from poverty, for others, wealth, religion, abuse, or simply the monotony of middle-class suburbia.

I started photographing again around this time, my attention focusing on what Warhol termed "lower bohemia." My photographs were black and white, all developed in high contrast and set in the streets, parks, and clubs of New York. As an artist and an icon, Warhol fascinated me. I believed his work held up a mirror to the shallowness and emptiness of the mass-produced commercial America, and this was the world that I perceived myself to be fighting against. Anything clean and commercial and corporate was bad. When Kmart moved into Astor Place, we protested. When TCBY opened up on Second Avenue and Tenth Street, we refused to buy our ice cream there, walking the four extra blocks to the punk-owned ice cream shop on Avenue A. When Giuliani began Operation Condor and created the Street Crime Unit, we marched against it. We didn't want a clean, crime-free village. We wouldn't wear Nike or Gap because they used sweatshops, and we snubbed new clothing, only secondhand was allowed.

I read voraciously about Warhol and the pop artists in those years. The floor at the foot of my bed was stacked with piles of books on Edie Sedgwick, Nico, Viva, Bridgett Polk, Basquiat, Raushenberg, Jasper Johns, Paul Morrisey, Billy Zane, and the Velvet Underground. Their world mesmerized me. It was exactly what I had been searching for all these years: a place where anything was accepted, where insanity and talent and guts coalesced to form this bursting, throbbing ball of human energy.

And I held Warhol in such high regard for facilitating this world that I didn't want to admit what I had always, somewhere veiled in the folds of my mind, suspected—that he hadn't actually created his art with intended meaning or statement, that it all happened inadvertently. Warhol himself had made statements about the vacancy of his work for years: "If you want to know all about Andy Warhol, just look at the surface of my paintings and films and me, and there I am. There's absolutely nothing behind it."

On some level, I was afraid to admit this to myself because I knew I'd then have to admit the same thing about my own life and everything I ostensibly stood for. To the naked eye it may have seemed that I had my values and beliefs in the right place, that I knew precisely what I was fighting for, but the truth was I didn't really have a firm grip on any of it. I was a flailing wannabe revolutionary floating along in a blind miasma of thoughts, ideas, and fears. Nothing was tangible or succinct or rooted. The way I lived lacked heart and sincerity. I was acting. I was full of contradiction. I was posing, affecting. It was all a performance.

27

CLOSE ENOUGH TO LOVE

I didn't see much of my parents anymore. They didn't approve of me or the people I surrounded myself with. They just told me how unsafe my friends were—such desperate, dirty people. They didn't like my photographs either. My father squinted like a sick child when I showed him my first portfolio. He called the photographs trash. That was the first time I allowed myself to hate him. But I was so still hungry for his love and approval that I always kept returning, hoping each time his reaction would be different. But he never failed to devastate me.

Occasionally my mother would call to check in. We'd chat for a little bit. She'd tell me I was destroying my life, and I'd tell her she was a pathetic socialite who knew nothing about life. Then her voice would turn into a sad hush and she'd say *goodnight dear* and I'd hang the receiver up and cry.

Once in awhile, they'd request that I join them for a Weaver "family" dinner, and I'd usually be amenable. The dinners were always the same: monotonous chatter, catching up with nephews and half-sisters I hadn't seen in years, sneaking off for cigarettes when the adults weren't looking. Only one night stands out from the tedium of these evenings.

I had snorted a few lines of cocaine before leaving my apartment, and all through cocktails I was high and chatty. Before we left my parent's apartment for the dinner I popped into the guest bathroom, which

smelled of potpourri and musk as it always had, and snorted a few more
bumps directly off my hand. The first elevator carried my mother, father,
two half-sisters, and myself. My mother was telling a long story that
didn't have much of a point other than to fill the silence. *Rosemary*, my
father finally said, *nobody is interested, would you just shut up.* Her usually
impervious smile shrank into her face. She swallowed, and I noticed her
eyes fill with tears. Years ago, she would have fought back, but now she
just stood there, watching the lighted numbers go down on the brass strip
above the door: *4, 3, 2, 1, Lobby.* My father walked ahead with my half
sister, my mother lagged behind, struggling with the velvet-covered buttons of her coat.

"This button, Ali. I think it's too big for the slit," she said.

"Here, let me try," I said.

I buttoned her coat, slipped my arm through hers, and we walked the
two blocks together as I imagined mothers and daughters should. I don't
know what came over me that evening. I imagine it was the cocaine, the
buildup of dopamine that generates continuous stimulation and joy. But
whatever it was, it let me sympathize with my mother and forget the bit-
terness and hurt for a moment.

The windows of the restaurant looked welcoming that night, with
candles and long-stemmed paper whites in wooden boxes. I peered in and
saw the others, slightly tipsy from the cocktail hour of champagne and
blinis, fighting over who would sit where, and my father, who was never
good at discipline, trying to quiet them all. As my mother and I stood
there in the cold watching, she leaned in and gave me a moist peck on the
cheek, exuding a scent of coriander perfume. And I took this gesture as
her way of acknowledging a shared union of unspoken discontent between us. Standing outside the window together, I felt she understood
what I had been running from all these years.

At the end of the dinner she turned to me as I hailed a cab and said
Ali, this is the nicest time I've had with you in years. That night was the first
time since I was a very young child that I felt anything close to love for
my mother.

28

THIS IS TO
MOTHER YOU

I quit my job at Saint Thompsons. I began to feel lackluster and obtuse in my paint-covered smock, cutting out hearts and stars, tracing letters, and wiping mouths and pouring juice for seven hours every day. I could feel my mind atrophying. And some of the teachers had begun to look at me strangely. I decided I'd better leave before their suspicions were confirmed.

Mikey hardly got work anymore because he was always oversleeping for jobs or nodding off on shoots, so for weeks we lay in bed all day, getting out from under the covers only when we needed more Ketamine.

The Velvet Underground played incessantly on the stereo, perhaps because we hoped to gain some innate knowledge from Lou Reed's lyrics, magical words that would free us all overnight, turn us into the painters or writers or singers we wanted to be. But the problem was we didn't do much to further our careers.

The extent of my attempt at painting was Mikey and I, high on Rohypnol, drawing long and emaciated dancing skeletal figures on canvas after canvas. Our brushstrokes were weak and barely able to graze the canvas, which gave the bones a featherlike texture. The figures seemed like they might shatter right off the canvas if you shook it too hard. We made loads of them and hung them unevenly around the apartment, blue on black, green on yellow, purple on black, orange on gray. On cocaine

From a self-portrait series

we were the most productive, collaging walls with underdeveloped black-and-white photographs and tinfoil, painting murals on each other's naked bodies, dyeing our hair a myriad of colors. We spent hours and hours writing screenplays or having photo shoots; one night I went through thirty-three rolls of film. Then there was our obsession with changing the wall colors every few months. We'd have painting parties where everyone had to bring a bottle of wine and a jar of paint. The walls were streaked and uneven, but we didn't care. We could barely see them most of the time.

During Christmas our apartment was lit up like a suburban lawn: flashing Santas, bright twinkling lights, reindeers and sleighs, two miniature trees with multicolored snowflake-shaped lights and fashion magazines cutouts as ornaments. When I was a child sitting in the backseat of our car, I remember passing the flickering homes of suburban Connecticut: brightly lit plastic reindeer in the front lawn, wreaths on every window, and lights covering the shrubbery out front—I'd stare through the windows in awe of the children inside those houses. But when I asked my mother if we could decorate our house that way, she laughed at me as if

she knew something that I didn't and said *I don't think so, dear.* Astro and
Mikey knew this story, and every December they'd spend hours turning
my apartment into a Christmas fantasyland, and when they finished,
Mikey would stand up tall, taking it all in with his hand on his hip and
say *Look at your daughter now, Mrs. Rosemary Weaver* as if he were somehow punishing her and getting something back for me, which I suppose
in a sense he was.

Falling asleep in my dirty black-and-white leopard sheets, sandwiched between Mikey and Astro, was far better than falling asleep in my
single, clean and ironed, blue-and-white bed at my parent's home.

Jenny introduced me to shooting during the spring of 1999. She carried
needles in her clear plastic Hello Kitty makeup bag. I saw them in there,
lying around with a few glittery pineapple-flavored lipsticks and sharpened
blue eyeliners. She took it with her whenever she used our bathroom, as if
she needed to reapply her lipstick every hour or so. But we all knew she
shot up in there. Ever since she'd started dating Little Joel, she'd begun
shooting the Ketamine like Joel shot his heroin. She'd also gotten a tattoo
of his name on the left side of her neck with a heart balloon on either side.
When Jenny died two years later, her mother insisted that the tattoo be
covered with foundation for the funeral because she didn't want to be reminded of Joel. It was, after all, Joel who got her addicted to heroin.

It happened quickly. It always did for me. The White Rabbit didn't
have time for baby steps. One day Jenny offered me a clean needle and
said I wouldn't regret it. She said the shooting high blew the snorting
high away.

"Don't shoot it in your vein, though," she said. "The drug is too potent, and you could die immediately."

We sat on my bed, side by side, our backs against the wall. She
opened her purse, and a gust of fruit and lavender entered the air. She
pulled out a needle still sealed in its wrapper and a glass bottle of Ketamine, known to us as a "lick," short for "liquid." She filled up the needle
and handed it to me.

"Roll your sleeve up 'cause you don't want to get blood on that shirt, and then just jab," she said. "The harder the better. Less painful."

"Just jab," I said. "Just like that?"

"Well, yeah," she said.

And I did.

By the following Wednesday, I had become a member of the Avenue C needle exchange. The tips of needles grew dull quickly, usually after three or four shots, and the duller the needle the more painful the experience, the more blood that poured from your arm, and the more likely for it to be black and blue the following day. Astro reprimanded me in his most fatherly tone when he found out I had been shooting with Jenny, but within a few weeks he was doing it too, and so was Mikey.

Jenny was right. The shooting high took the drug's potency to another level. The high came over me in the gentlest of ways, like I imagined death would come. I could feel the liquid distributing itself throughout my body in a blanket of mild tingles. My eyes would grow foggy, my limbs heavy and weighted, thoughts would cease, movement stop, and I would gradually be erased from the world.

Together we'd line up outside the needle exchange holding brown paper bags filled with old needles, a bottle of Ketamine tucked away somewhere in a pocket or shoe. The line snaked around the corner of the graffiti-sprayed tenement building, fifteen or twenty junkies always in line, dragging their sandbagged feet across the pavement of spilled soda and dog urine. Most of the buildings on the block were empty storefronts with windows covered in a thick layer of film, some shattered into fang-like teeth allowing us to peek inside, where teenage squatters slept on the floor.

There was nothing like the feeling of getting home with an entire bottle of Ketamine and ten brand new needles sealed in plastic that would slide smoothly into and out of my flesh without that blood bubble appearing, or that crunch of skin dull needles always produced. It was a feeling of contentment and safety, like what I imagined chipmunks must feel after they'd gathered enough acorns to settle in for the night.

Amphetamines made me think and worry and plan too much. They shined a large bright strobe light on all that was wrong with the world, and I'd spend my nights in heated debates over this and that, always the same issues: the Columbine shooting, racial discrimination, the Amadou Diallo case, poverty, immigration, the death penalty. And on and on we'd go, talking these issues raw, and when we'd said all that could be said five times, we'd say it again because it was the only thing that ameliorated their existence even slightly.

On Ketamine, I spent my days and nights ritualistically soaking alone in a bathtub filled with aromatherapy bubbles while listening to Sinead O'Connor sing "This Is to Mother You" on repeat. I'd bring my journal into the tub with me and scrawl out some morbid poetry as my vision blurred and my mind was washed with shadows. Then I'd lie there in a mist of detachment for hours and hours until the drug wore off and the bubbles popped and my fingertips were lined like corduroy. Alone with the water lapping the sides of my bare body, my most private thoughts recorded on the page, and Ketamine cradling me in its warm blanket of numb safety was the closest thing to comfort and security that I felt in those years. If someone had sliced me down the middle, they would have found no life, just the remainder of rotting, bloodless organs like dried jellyfish on a once-inhabited shore. I made a trade that seemed

Nodding out in the East Village studio

fair at the time: my soul for deadness, a space of white nothingness in which to exist. And it worked for a while. But then the trade malfunctioned and my sensations began to return so frequently that I couldn't keep up. They were like hundreds of monstrous regenerating cancers, and though I worked in fast-forward to keep myself void of spiritual and emotional depth, I could only cripple it, deform it slightly. And what is even more painful than complete self-destruction is severing only a part of yourself and being left with that sliver of life that you can't quite abandon, no matter how hard you try.

29.

DESPERATION

There must have been over a hundred people crammed into my little studio that night. People sitting on top of each other, people huddled in groups passing joints or pipes filled with meth or PCP. A couple was having sex in the walk-in closet. I sat propped up in the corner of my bed with a fluffy zebra print pillow on either side of me. I was on just the right amount of coke and Ketamine and whatever other pills I had taken. People came and plopped themselves down beside me and I watched as their mouths moved and noises came out of them.

I could see Mikey dancing in the center of the room. He had his expensive sheer green shirt on and his tight bell-bottom Costume National pants. The shirt had two prominent bloodstains on the upper right arm. As he danced, he blew kisses at no one in particular. His eyes swarmed with bright red capillaries and dark bags hung heavily below, while pale flesh flapped under his arms like the jowls of an asthmatic old man. He hadn't gone to the gym in months. Beside me was a stack of magazines he'd recently been in. I picked one up and flipped through it. Mikey, in a pinstripe blazer and pale blue open-collared shirt with clean, white skin, clear eyes and plump lips red with life, beamed back at me. Then I returned my glance to the real live Mikey dancing in the center of my apartment as Donna Summer sang *I feel love, I feel love, I feel love,* and I thought of Dorian Gray and decided he must have confused it. It was the magazine pictures that were supposed to bear the burden of age and infamy, not the living, breathing person.

Partying with a friend

Astro had nodded out in the bubble chair. His head rested against the unevenly painted blue wall; his eyes and mouth remained open and drool trickled slowly down his chin like a melting ice cube on the corner of a table. I watched a group of girls with Cosmos in their hands walk by whispering and laughing at him. I wanted to get up and smack them but I didn't have the energy.

Our drug dealer was there. This was the first of our parties that he ever attended. He stood alone in the corner with a baseball hat that left triangular shadows across his thin chin. There was something about his timidity, the way his mouth drooped to the side when he smiled, the way he cleared his throat before speaking, that forced you to believe there was something important inside him.

Strange faces were there, too, people who only went out on weekend nights, milder users who just dabbled in drugs. We knew they looked down on us. We knew they frowned on our filthy, darkly lit apartments and raccoon eyes, but we also knew that a hidden part of them couldn't help but respect the way we gave ourselves entirely over to the other side. And when I was high I almost liked them and they almost liked me. They'd scan our studio with beady, blinking eyes absorbing the strange way people lived when they stopped caring. I could tell they were oddly attracted to our dark world, the way people are attracted to bloody car wrecks and junkie models. One or two of them would always get lost in it, and it wouldn't be until sensing their friends' fear and disapproval that they'd realize they should be scared too. One girl, whose name escapes me now, would ask me to mix her tapes of music, but we had an unspoken agreement that no one else was to know about this exchange. She also asked me to shoot her up once. She said she wanted to try a speedball, a combination of heroin and cocaine, but she was too frightened of needles to do it to herself. I took her into the bathroom and jabbed it into her arm and I didn't feel bad about it. I knew I should have felt bad, but I felt like I'd really helped her out.

Later that night, after the crowd had thinned out, I was lying under the covers, high on Ketamine, and I felt a hand between my legs. My eyes cracked open. I made out a blurred man. I recognized him. He was a club dealer. We'd never spoken. He slipped his rough hand under my halter top and played gently with the hard tips of my breasts. His other hand was still between my legs. I knew this wasn't exactly right, but I didn't care. I closed my eyes and let him touch me. I was flattered to be wanted so much that the person who wanted me believed they had to sneak in to get it, like I was so precious that he'd never be able to have me if I were awake.

I just lay there enjoying it, feeling his dick harden next to my thigh. I felt him push my tiny skirt up to my waist, slide my skimpy lace underwear down to my ankles and spread my thighs. I felt him lift himself up with one hand and his body seemed to levitate above me. I heard a zipper pulled down and felt the thick, stretched skin of dick against my leg. I was coming back a little now, being pulled out of my haze of disconnect. I was beginning to realize the magnitude of what was occurring. I was beginning to realize it wasn't just about the momentary sensation this man's hand was giving me as it rubbed itself between my legs.

The next thing I remember was someone yanking him off me and punching him and screaming, "Get the fuck out! Hetero rapist!" It was Astro. I heard the door slam and I sat up, messy and barely clothed. Astro looked back at me, really looked at me, with his hard, exploding eyes, and

I thought he knew. I thought he saw the despair inside me that had allowed this to happen. I thought he saw where my hopeless craving for love and affection had gotten me. Our eyes clicked into each other, and I realized he *did* know and that knowing had finally allowed him to understand me.

"Are you okay, L'Ali?" he asked.

"Yeah, I'm okay," I said. "Can we get these people out of here? I think I need to go to bed."

Astro directed the remaining stragglers out of the apartment, filled me up a needle of Ketamine, and said:

"Here, do this. It will take the edge off."

"Edge off what?" I asked.

"Whatever," he said.

The first time I shared a needle with Mikey it was sleeting against our box-shaped window. I could hear the trash in the alley whirring six floors below. We were painting over a used canvas because we didn't have money for a new one.

"Pass me the blue tube," I said, gesturing across him.

Astro opened the door.

"I have an eight ball of cocaine and one liquid. Preferences?" he said.

"I'm assisting a photo shoot tomorrow for David LaChapelle," Mikey said.

This was his way of telling us he had no money to contribute.

"Blah, blah, blah," Astro droned, rolling his eyes.

"We have no clean needles," I said.

Mikey and I spent the next hour searching the Lower East Side for needles, but no one was out. We looked in all the usual spots, by the chess tables in Tompkins Square, the punk's benches, in front of the A and P on Avenue C, the homeless shelter, the corner of Second and Avenue D, under the basketball hoops, and outside the needle exchange.

The night was warming up and the sleet had turned into a mild drizzle. I could feel a layer of precipitation forming across my skin as cars

sped along Avenue A kicking up city trash and spitting it at our ankles.
My stomach hurt. The few people we knew were out of needles, so we sat
cross-legged on a bench in the park and waited. Within minutes it began
to pour. We got up and walked the route again and again, hoping a new
face with fresh needles would appear but no one did. The pajamas I'd
had on for the last three days were soaked and hung heavily on my body
and my stomachache had become one long, sharp pain.

"We should have just gone to a Duane Reade and gotten bleach," I finally said.

"Let's go now," Mikey said.

"Won't be open now. It's past midnight."

"Don't they have a twenty-four hour deal? This is fucking New
York," he said.

"You know, I might have an old needle in one of my bags. Let's just
go back. What else can we do? We've been searching for two hours,"
I said.

When we arrived home, Astro was already high. He'd cooked part of
the bottle and now sat stiffly on the chair staring through the television
set. In the perpetual neon green glow that was ever-present on *The
X-Files*, Agents Mulder and Scully were scrutinizing some washed-out
photograph of abandoned city streets filled with half-human hovering
figures. Astro couldn't speak, didn't even know we had returned. I grabbed
the bottle from his loose grasp and handed it to Mikey.

"You go first while I look for another needle," I said.

But I didn't find another needle because there wasn't one to be found.

Within minutes, Mikey was crouched on the floor, frog-like, in his
tight black Helmut Lang pants and gray Hanes T-shirt. He seemed to
be staring at me, but he wasn't. On Ketamine, you were never staring
at anyone or anything concrete but rather looking past it, into the so-
lar system. He began to crawl around swinging his arms in circles,
muttering something about becoming fabulous. Then he got quiet,
placed his hands flat upon the prewar floorboards and crawled toward
the futon mattress, where he fell into a fetal position half-on, half-off
the bed, drooling.

A weird sadness lingered in the air that evening as the sandalwood incense burned from the windowsill. Two bare blue lightbulbs hung from the ceiling, giving off a tungsten glow. We'd changed the bulbs months ago but never bothered to put the glass covering back on. I hated myself for what I was about to do, hated myself for hating myself so much that I would allow myself to do this, but I needed to get high. My stomach had developed an ulcer from the Ketamine, and when I wasn't on the drug I could barely stand up straight because the pain was so severe. I could feel my body turning itself inside out. I was sweaty, cold and then hot, hot and then cold, and I felt like my organs were going to jump from my skin.

I did my best to convince myself one could only catch HIV from shooting intravenously, though I knew blood was involved either way, and wasn't blood the problem? I wasn't sure. I fought with myself for a while. I tried lying down but couldn't sleep. I tried having some cereal, but my hands were shaking so badly that I spilled milk across the counter. Finally I walked over to our dresser and lifted the needle from where Mikey had left it. I carried it to the sink and stuck it under the faucet of hot running water. I inserted the metal needle into the gray rubber top of the lick, pulled the white plastic stopper back *slowly* to limit the formation of air bubbles. I held the needle vertically and tapped it to destroy any accidental bubbles. I felt around my upper arm trying to find the area that ached least with black and blues, then I pinched my flesh and jabbed and some blood came spurting out because the needle was old, used maybe twenty times that day, and Astro suddenly came back to consciousness, and he stared at me, and as I left the world he said, pointing at the red blood design on my T-shirt sleeve, *It's art, L'Ali, you made great art.*

Six months later I found out Mikey was HIV positive. He showed up at my studio one day. His hair was oily and sticking to the sides of his forehead in perfect curls of dirty sweat and summer humidity, his cheeks were flushed, and his eyes bugged like he'd just seen a hundred bludgeoned corpses. We'd fought badly a few days earlier. Maybe he'd stolen money from me again. I'd kicked him out for the fifth or sixth time that month.

"I need to talk to you," he said, as he stood in the doorway.

"You're not welcome here anymore," I said. "I can't keep doing this."

"I don't want to stay. I just need to talk. I have to tell you something. Please let me in, just for a minute, and then I'll leave."

I let him in and he curled up on the sofa next to me, his big New Balance sneakers hanging off the end. He pressed his fingertips into his eyes, and I thought he might be crying, but you could never tell with Mikey. He was always acting.

"I just came from the doctor. I got tested, and it came out positive."

"What? Are you sure?"

"Yes."

"But sometimes those tests are wrong. I've heard that you can test positive here and fly to Australia and test negative," I said, panicking.

"Well, are you going to fly me to Australia?"

"No, but I'm just saying the tests aren't very accurate."

"I can go back in six months and get another one, but I think it's pretty accurate," he said.

He had his head on my lap now and he was sobbing real tears. His messy, soiled black hair spread across my bare thigh like pine needles covering a lake, and I rubbed the wet back of his navy tank top. He smelled, and dirt encrusted the creases of his neck. I could tell he hadn't bathed in days. I didn't know what to say. I didn't know what to do with him.

"You should go and get tested," he finally said, "just in case."

"I guess," I said.

But I didn't. It was too terrifying, and somehow my warped thinking led me to believe that it would be better not to know at all. I figured if I got sick I'd know. It wasn't until years later that I actually got up the nerve to be tested and turned out to be negative.

Marilyn wrote a song about me that summer. One day I was over at her apartment on Twenty-second Street, and we were passing a joint back and forth as I was talking her out of checking herself into rehab for the third time. Lou Reed's *Transformer* album finished, and the five-disc CD player clicked into the next slot. It was her voice.

"Is this you?" I asked.

"Yeah, I've been recording again. Gonna send it out to a couple people."

"Nice," I said, as the song began.

Clean works are a dollar but the price is going up.
I'm having trouble breathing but I don't think you believe that.
Just forget about it all, flying circles into walls.
You don't look quite so fabulous with bruises on your arms.
Bleach is such a hassle it just takes way too long.
Lock yourself in that stall you ain't foolin' us at all.

"Is this about me?" I asked, startled.

"Ah, yeah, remember," she said. "I wrote one about you."

"Vaguely," I said. "It's fucking depressing."

"Yeah, well, I wrote in rehab," she said. "You know how you get in rehab."

Don't know if I can meet you downtown.
It will be too hard to stick around.
Traffic gets too heavy when I'm downtown.
I'll look for you in lost and found.
Your shirts are stained with blood.
You left some on mine too.
Just keep flying till you fall, I'm just waiting for that call.

"I have to get going. LaChapelle has his opening tonight at the Toni Shafrazi gallery. Do you want to come?" I asked her.

"Nah, I'd rather just lie here and listen to Lou."

"Okay, see ya," I said.

"Hey, remember," she said, smiling. "If you want to sing out—"

"Sing out," I replied as I left, shutting the door quietly behind me.

30

BECAUSE THEY
WERE REAL

During those years after Cascade, the only thing I cared enough to stay sober for was my photography. For me, it was just another escape. Inside my rented box-sized dark room on West Seventeenth Street, I was able to sever all connection to the outside world. I liked myself inside that tiny room, smelling of sodium hydroxide, borax, and metol. I was at work on a series of portraits, each one attempting to capture the thread of tenderness behind the terrifying exterior of many of my subjects.

At twenty-one, I had a show in the gallery of a popular club I frequented. Robert Roberts was the gallery host. He was albino-pale and perilously thin with large, tubular eyes, a light covering of red-blond wisps across his head and a neck that could rotate 230 degrees. He said he was nocturnal like a bat or an owl, which was why we never saw him during the day. He liked me. He introduced me to important people I knew I should have cared about meeting: hotshot photographers like Terry Richardson and David LaChappelle and Cindy Sherman, and other big names like Donatella Versace and James King and Marilyn Manson, but all I wanted to do was excuse myself and get high in the bathroom stall.

I invited my parents to the opening night exhibit, and they came in a navy blue limo, my mother wearing her ostrich fur coat and my father in a tuxedo. They were coming from a black tie benefit. The club had recently partnered with PETA and didn't allow fur inside the doors, but my

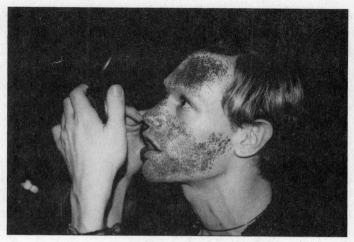

A portrait of Richie Rich from the show

mother told the bouncer her coat was fake, and he believed her. I saw my father's face emerge from the crowd of high-fashioned patrons in provocative dresses and designer sunglasses, drag queens named Jem Jender or Shimmer, Glimmer, and Glow and men in tight T-shirts and eyeliner. He looked simple and elegant, a relic of a time long past. I welcomed him and asked if I could get him some water, but he said, *Get me out of here. Your mother made me come. I came, now get me home.* The music thumped and pounded from the speakers around us. It was loud and technolike, and with every sharp beat I saw him squint his eyes like he'd squinted at my photographs. He said the people were ugly and that my pictures scared him, so I walked him to the limo waiting outside and kissed him goodbye.

Then I stood outside the club smoking a cigarette.

I watched a famous photographer with dark glasses and a purple pinstripe blazer get out of a white limo followed by his most recent muse. She was naked with huge painted tits and red tassels hanging from her nipples. On her crotch was a small triangular piece of red patent leather held on by what appeared to be cinnamon-flavored dental floss. Her ass and her lips were blown up like a cheap plastic doll and seemed to carry her weightlessly through the crowd. *Looking good*, I heard someone say.

Fuck off, replied the photographer, shielding her from the crowd. I looked up at the night. It had begun to snow. Silver flecks darted from the dirty sky like a line of bullets in rapid-fire, merging into the funnel of light that poured from the street lamp above me. It looked so warm and safe, that streak of yellow light with its soft silver bullets falling in succession, one after the other, until melting into the muddy sidewalk below my stilettoed feet. For a minute I imagined the streak of light to be heaven beckoning me upwards and I nodded, asking to be taken. "Alison," I heard. The heavy black side door opened and Robert Roberts peeked his small head out. "Aren't you coming in, baby? It's your night."

"I'll be right in," I said. "Just smoking." I lifted my hand where a half-smoked cigarette with a long, gray ash rested between my pointer and middle fingers.

"Okay," he said, shutting the door as his big, bulbous eyes blinked away snowflakes.

It dawned on me, as I stood outside that club in the drifting snow, that instead of walking my father to the limo and kissing him goodbye, what I had really wanted to do was lock him up in there with all those clown-faced freaks and the loud thumping music and my scary photographs. I wanted to force his eyes open so he could really see what was in front of him. Not the ugliness and horror of it but the desperation and shame that was underneath. I wanted to make him understand that desperation and shame could be beautiful too, beautiful because it was as real as a bleeding, stuttering, pulsing heart.

My mother, on the other hand, stayed into the night, dancing with my friends and being her charming self. I watched her from my spot in the corner of the banquette, smiling and sipping champagne, dancing in circles with one arm raised and waving as if she were Jackie O. She flashed before me as the young woman I'd seen in pictures now stored away in shoeboxes.

In one of those pictures she stands to the side of the United Airlines logo in a white button-down shirt with a seventies collar. She's maybe in her late twenties. Her pale lips are parted and smiling, and she has been in the sun and has a healthy tan, and her hair is blonder than usual and pulled into a ponytail. Gold hoops dangle from her ears. She seems content to stand there next to the logo, almost as if she knows it won't be hers for much longer. My mother's smile is strong and confident, nothing like the weak, quivering smile it is now. She is the director of United Airlines Public Relations. She caters to clients like Redford and Newman, as she calls them when telling her stories years later.

As my mother partied into the night, she exuded a confidence and energy I had never seen in her. Immersed in a world other than her own, she was exuberant and whole, lit up like a freshly purchased lightbulb. My friends took to her immediately, and I sat back, content to watch. And it was then, for the first time, that she made sense to me. There in the dark, overcrowded club watching smoke curl eel-like from the tip of her cigarette and light up purple in the neon club illumination, I saw the woman in that old picture, a woman without constraints, unsaid rules to uphold, or the position of society wife to fill. And for years afterward, I would get glimpses of her late at night in Connecticut, laughing by the fire when she felt safe with the company, when she felt no one was looking or that no one would tell. In those moments, a woman I recognized as authentically my mother would appear.

31

A MOTHER'S WOMB

Around my twenty-second birthday, my mother began to suspect me of using again. One night she called around ten or eleven in the evening to check in on me, and we had a strained conversation, both of us trying to fool the other into believing the lie.

"Things are good then?" she asked.

"Yeah," I said.

"We should have lunch one of these days."

"Sure, maybe next week."

"Are you still living with those people?"

"Yes, they're my friends, Mom."

"You know your father misses you," she said.

"Yeah?"

"Yes."

"He could call," I said.

"You could, too."

"I have to be at class early. I've got to go to bed."

"Okay, sweetheart, goodnight."

"Goodnight, Mom," I said.

Then I hung up and ran a bath, shot some Ketamine into my arm and soaked, half-dead in the rose-scented water, eyes rolling back in my head, pulse and heart sluggish. I fell asleep in the tub that night, or over-dosed. I am not sure which. The door was locked. From the radio, Steven Morrissey of the Smiths sang a slow and weary dirge about falling gently, eternally, into sleep. Astro was pounding on the door with his fists but I didn't hear him. The bubbles deflated. The water in the tub grew cold. Astro called a locksmith.

I woke up to him kneeling by the side of the tub, shaking me and screaming. "Never lock the fucking door again. I thought you were dead, L'Ali! I thought you were fucking dead!"

"I'm not dead," I said weakly.

He lifted the empty Ketamine bottle from the side of the tub.

"How much of this did you shoot?"

"I don't know," I said.

He carried me from the bathtub, dried me off, and slipped me into bed.

"Go to sleep, L'Ali," he said.

And I did, the low rumble of Morrissey's voice carrying me off into the weary dark.

One day during the summer of 1999 it all blew up in my face. It was sweltering in the city, painfully hot and sunny. Mikey and I spent our days nodding out on a shady blanket in Tompkins Square Park with Jenny and Little Joel and some park squatters. Sometimes we went to the public pool on Carmine Street. It was a Friday, and we stood in line with thirty or forty school children on a field trip, Mikey in his tight black Helmut Lang tank top, cutoff jeans, and Elizabeth Taylor sunglasses, and I in a multicolored seventies thrift shop sundress with makeup from last night's party smeared across my face and glittering in the sunlight. We hadn't slept yet.

The day was hazy with unmoving smog, and the hot air weighed on me like the leaden radiation shielding blankets draped across you in an X-ray room. I walked through the door that said "ladies," and Mikey veered left, pushed along by rambunctious boys slapping each other around. I locked my things in a top locker, clipped the orange key onto the side of my bathing suit as usual and went into a bathroom stall to shoot up. I thought I had my timing down perfectly. I knew after I shot the K, I'd have about four minutes before total incapacitation. I could apply sun lotion and get out of the changing room within two minutes, dunk in the pool for one minute, and be safely on my towel before I nodded out completely.

I remember my feet breaking the water on the steps. My big toe seemed to hold an almighty Neptunesque power over that water, capable of shattering the sheer flatness of it again and again as I moved further down the steps into the pool. After that, all I remember are senses, moments of texture or smell I can't quite define: the smoothness of the rippled water gliding across my fingers, the smell of chlorine on scalding hot cement, baby's-pink sun block, a substance of soft diluted glue against my chest, my neck, my face breaking the water, my body falling weightlessly into safety and silence like reentering a mother's womb. And then— blankness.

I don't remember what happened after that, but Mikey said I looked like a blue goddess rising from the sea when the policeman pulled me out of the water. He said blue eye shadow was running down my face, and my hair seemed to wrap around my shoulders, clinging, a protective cloak. And the thin, wet layer of pink sun lotion shone like embryonic fluid as the police officer lifted my limp body into the air. *You matched beautifully. You wore a blue-flowered bikini. It was almost like you knew you were going to be the spectacle of the Carmine public pool. You definitely dressed for it. I'll never forget it. Blue, blue everywhere; even your earrings had a blue stone in them:*

When I came to, I was chained to the metal fence that separated the pool from the street. A boy of about seven or eight in yellow bathing trunks jumped in circles on the diving board in front of me. We made eye contact as his feet pushed upward and his body hung in midair. He opened his mouth wide and screamed, *junkie,* with an accusatory finger pointed my way, as he turned and cannonballed into the pool. I'll never forget the way his lips came together to form that word.

I was taken to the hospital in my blue-flowered bikini, given a hospital gown and all sorts of tests, and then sent on to jail, where I spent the night in a private cell in the "sick unit" because the police officer had lost my hospital papers and they thought I might be carrying a contagious disease.

There were no beds in the cell, just a wooden bench, but that was fine with me because it felt less permanent. I lay on the bench and picked long flimsy splinters from its underside. I pulled my legs as far into my ass as they would go. My bathing suit was still damp and it felt cold against my skin. I tried to cover myself with the hospital gown that hung around my body, but its napkinlike consistency provided no warmth. In the cell next to me I could hear a woman talking to the guards on either end of the corridor. She was screaming about her husband and the president of the United States. She was letting them know that her husband worked as an undercover spy and that they would be sorry for what they'd done, and she believed it too, every word of it. You could hear it in the strength of her shrieks.

When I saw the thin lines of light penetrating the sides of the small covered windows on the cement wall opposite the cells, I began to worry that nobody was coming for me as they'd promised the night before.

"Hello?" I said.

Nobody answered.

"Hello?" I yelled. "Hello, is anybody there?"

I couldn't see the guards anymore and the door closest to me was firmly shut.

"They can't hear you," said the woman whose husband was a spy.

"I'm supposed to be going to court this morning," I told her.

"That's what they told me two and half months ago."

"You've been here for two and a half months?" I asked.

"Well, they think I have," she said. "Sometimes I sneak out."

I stopped listening to her and began to rattle the bars and scream for the police, but nobody came. I lay back down on the bench and the woman said, "Told ya so."

A few hours later a guard showed up on the other side of my cage and handed me a brown paper bag. In it was a ham and cheese sandwich and a child's-size Tropicana O.J.

"Eat up," he said. "We're taking you downtown in ten minutes."

He walked by me toward the woman's cell.

"Are you ready, ma'am?" he asked.

"Don't try and trick me," she yelled. "I know what you're up to."

"We're taking you to court."

"Where's my husband? I ain't buyin' this shit from you," she screamed.

Two other police officers entered, unlocked her cell door, and escorted her down the corridor. I was standing at the front of my cell washing down my sandwich with the juice. He looked at me and smirked.

"Lady is nuts," he said.

"Well, she's been here for two months. She never thought she'd get out," I said.

"Two months? We brought her in last night, just a few hours before you got here."

Ten minutes later, the woman and I sat together in the backseat of the cop car zooming down Fifth Avenue. She was screaming through the metal divider at the two cops in the front seat, her long, thin ebony fingers clutching the chain links with great determination. I stared out the window into the sunny summer New York day and saw a woman who could have easily been my mother trotting down the street in a loose blouse and large circular sunglasses with yellow lenses and clear frames. A bulky leather bag hung off her shoulder. She waved at another woman, this one in a yellow brocade suit, who stood under the navy awning of a private club. Once she reached her, they kissed on both cheeks and disappeared into the safety of the building.

The woman beside me quieted and leaned toward me.

"Fucking assholes—these cops," she said. "They won't even acknowledge me. They won't even give me the respect I deserve."

"Fucking assholes," I said in my toughest voice, nodding in agreement with her.

She patted my thigh and started screaming at them again.

I felt a great camaraderie between the two of us that morning, like we'd fought in a war, watched our brothers bleed to death before our eyes, and lived to talk about it. But of course we hadn't. We hadn't done anything admirable at all. And what I failed to recognize then was that this was real for her, that the dirty cell in the sick unit was real for her, that the *sickness* was real for her. There was no reconciliation or healing ahead, no family to bail her out. This was the stuff of her life.

Four hours later, a judge sentenced me to three mandatory court rehabilitation sessions, a fine of three hundred dollars, and fifty hours of community service.

And it was then that I knew things had to change.

The thing is, addiction is a very ambiguous topic for me, and I've never been sure how to go about discussing my recovery. Mine was a backwards, tortuous journey. The type that certain sober people would tell you is the wrong way to go about getting clean. What would you say if I told you that I slipped up and did cocaine two summers ago? Would I still be recovered or would I be an addict? Certain sober people would say I was still an addict, and I would always be an addict. But I wouldn't listen.

I never went to rehab or joined the twelve-step program and went to meetings like Marilyn and many of my friends did. After Cascade, I wanted to be as far away from anything resembling a "program" as possible. I didn't believe that I had a disease, and I knew they would tell me I did. I knew it would become my life, and I'd have to deal with sponsors and confessions and key chains and anniversaries. I wanted to do it my own way.

My recovery occurred over many years. I moved out of my colorful drug den and into a white one bedroom on Astor Place. I tossed old phone numbers in the garbage. I let my hair return to its natural color. I quit Ketamine but picked up a mild cocaine habit again, justifying a one-night usage over and over again as the light crept in between the blinds, and I was forced to stare out the window at the real people and smell the nauseating aroma of sausage from my neighbor's stove, while I sat alone hating myself. I met a boy. He moved in and we called it love. I quit the coke but drank heavily. I felt empty. We broke up.

Like most things in my life, it was a slow and gradual process filled with unnecessary complication, but it was the way I needed to do it. Yes, I believe at one point I was an addict, but I am no longer, and these days I can confidently pick up a glass of wine without fearing that I'll be calling my dealer by the second one. That part of my life is over.

When I tell people this now, they don't seem to understand. They always think there should be something more, some AA program or religion or shrink who flew in and saved me. But this is it. One day I decided I wanted a real life. I struggled and I fought and I slowly won. There simply came a point when I realized it was no longer about making others pay, about punishing the people who forgot to love me along the way or about dwelling in the safe yet crippling zone of self-pity. And when it ceased to be about either of those things, I could finally move forward.

32.

CASCADE SURVIVORS

I often try to discern whether Cascade did me more harm or good. On one hand, it shed light and clarification on my past and gave me permission to break a silence that had concealed the truth about my family. For that I am grateful. But it also harmed me tremendously. The theoretical intentions of the Cascade School had been to save children from their addictions and depressions, but the process of recantation, self-criticism, public humiliation, and public apology, this *reformation of the character*, as they called it, was not only damaging, but dangerous.

Cascade graduates have been failing for years; some commit suicide, others die of drug overdoses, some just waste away. Recently one of our class presidents died, suicide by overdose; his name was Brett Lear. He was intelligent and kind and had a seven-year-old son. I can still remember the work he did in Forums. I can still see his face screaming at the floor in red-fire fury. In 1999, a year after Brett's death, his Cascade Big Brother died—another suicide by overdose. I didn't know him very well, but I remember his face too, concave cheeks and big swollen lips. He was handsome, effeminate, and smiley.

My thoughts and opinions surrounding the Cascade School will never be entirely clear. I've tried for years to define my relationship to it. I've tried to produce a tangible, assured, and judicious answer when posed with the question of whether Cascade was a good or bad place—as I so often am—but I am yet to succeed. There are moments when I hate and despise the school, moments when I miss it relentlessly. And moments when the easiest thing to do with it is laugh.

On January 20, 2004, the school closed down. I've heard many different reasons for the closure, but the most commonly voiced reason ultimately was financial. The students began protesting. They refused to comply with the rules. They smashed windows, vandalized the dorm walls, broke into the medical center, and ran away repeatedly. They fought more vigorously than we'd ever fought.

Cascade was not a lock-down facility and, ill-equipped to handle this behavior, it was forced to send the students to lock-ups or wilderness programs. Enrollment dropped as parents began pulling their children out. Eventually it was left with only thirty students and, struggling financially, was forced to close its doors.

Now all that remains of the Cascade School is an Internet chat group called the Cascade Survivors, consisting of ex-students who complain about the treatment they were forced to endure, or revel in the perversion of their Cascade memories. It was through this group that I learned about the history of the school. As it turned out, Cascade was an offshoot of the infamous Synanon cult that began as a drug rehabilitation program for adults. It was started by a man named Charles Dietrich in the late fifties, and received a lot of coverage in the late seventies due to two attempted murders of Synanon members who were trying to leave the cult.

According to the Survivors group, a wealthy man named Mel Wasserman spent time at Synanon, donated money to the school, and soon decided that he would start his own rehabilitation community for teenagers. He opened the doors to his Palm Springs house, inviting troubled children into his home to talk and calling the little community CEDU—a term that came from the philosophy of "see yourself as you are and do something about it." He used all of Synanon's methods and coercion tactics, and soon young kids were moving into his place. Taylor Gilbert, the Cascade school headmaster, was one of the men Mel hired as a "counselor," but once CEDU turned into a legitimate business with multiple programs under its guidance, all the early founders had to go. Taylor Gilbert moved to Whitmore, California, and started the Cascade School.

Like Cascade, most of the CEDU-philosophy schools are gone now, brought to their knees by a plethora of lawsuits and negative publicity. Two former residents of CEDU's Brown Oaks Treatment Center in Austin, Texas, alleged that they were sexually assaulted in 2002 by an Oaks employee, who ultimately pleaded guilty to assault charges, according to court papers. In 2002, a seventeen-year-old boy died at CEDU's Brown wilderness program in Texas after being restrained by camp staff members. State regulators said that staffers used improper restraints, but a grand jury handed down no criminal charges. Also that year, CEDU paid a $300,000 settlement to two former students after they were hurt in what students at the time describe—and the company confirmed—was a riot at CEDU's Northwest Academy in Bonners Ferry, Idaho.

On April 20, 2001, a former Cascade counselor named Julie Ponder was sentenced to sixteen years in prison for killing a ten-year-old girl named Candace Newmaker during a session of "rebirthing" therapy. Wrapping her seventy-pound body in a flannel sheet, piling eight pillows, and 673 pounds of body-weight on top of her, they instructed her to simulate birth by wiggling out of her flannel "womb," therefore becoming reborn to her adoptive mother. However, the adults did everything they could to frustrate her efforts to comply: blocking her movements, re-tying the ends of the sheet, shifting their weight, and ignoring her cries for help. In the videotaped incident shown in court, as Candace literally struggled and screamed for her life, protesting that she felt like she was going to die, they answered, "Go ahead and die." When she finally went silent—after nearly an hour of the suffocating therapy—the former Cascade counselor mocked her with the words, "Quitter, quitter, quitter. She wants to quit!"

Some members of the chat group are still consumed by their experiences there, whether remembered with nostalgia or loathing. They still speak in Cascade lingo or call each other by their old nicknames. They plan reunions; I attended one. It was in the bar of a seedy Las Vegas hotel, "The Circus." A woman with awkward, male features and smoky eyes collected ten-dollar bills at the entrance. Behind her, soft silver steam seeped slowly from fake crackling molten rocks over the rainbow colors of an electric disco ball. "It's all you can drink until midnight," she said and nodded in recognition at each familiar Cascade face that passed through the door.

Most everyone there was high on amphetamines. Some people had babies asleep back in their hotel rooms. They passed wrinkled wallet-size photographs around while mingling through the sweaty crowd. The rest exchanged quick glances of recall from people they had once slept, ate, and cried next to. The hopelessness of fear spread; the darkness had triumphed. It was clear that our two years at Cascade had done nothing to save us from our demons.

33.

HAPPY TURKEY DAY TO YOU, TOO

Eventually, on the way to sobriety and cleanliness, determined to get my life back on track, I resolved to recapture the carefree, fun-loving relationship my father and I had once owned flawlessly. The trouble was, my father was no longer there. In his place was a volatile, bitter, and depressed man who couldn't hear very well and dreaded human interaction. I am speaking of a man who once adored nothing more than being the center of attention, a man who embodied an energy and zest for life and the art of conversation like no one I have met since.

Now, he sat alone in his study reading, watching the news or the History Channel, sipping Ensure, and swallowing Paxil. I began to visit him once a week. I took him to the bookstore or for walks around the block. Some days, he was as strong as an ox, and I secretly believed he would soon return to his old self, but then there would be a stretch of days when he would list forward as he walked, and my hand would fly out in front of him, terrified he'd fall headfirst onto the pavement.

One afternoon he asked if we could go to the pharmacy and buy toenail scissors. He said my mother had hid his because she was afraid he'd cut himself, but she refused to touch his yellowing toes and shipped him off to the podiatrist every month. *The podiatrist is a real asshole. You can't trust a man that works with feet for a living*, my father would say. He was cheery that day. I bought him the nail scissors as he complained to the

The last Weaver Christmas card

clerk about the price and made me swear not to tell my mother. We walked home with linked arms, and when I kissed him goodbye in the living room I said, *I love you*, Dad, and he said, *Ditto Al*.

The first time I came to the house and my father refused to go out, it was spring, and he said it was too cold. His body slanted across the couch, covered in a maroon cashmere blanket, and only his slippered feet were visible. I sat down next to him and asked what he was reading. He didn't reply but lifted the cover to my face.

"Any good?" I asked.

"Nah, crap. I've read all the good ones twice already," he said.

"Are you feeling all right?" I asked.

"Lousy," he said, and I swear I saw tears welling in the corners of his eyes.

"What hurts?" I said, but I knew what hurt.

"Everything," he said, his voice trembling.

"Are you upset or sad, Dad?" I asked.

"Yes."

"About what?"

"Everything," he said again, his eyes now clearly filled with tears.

I wish I had taken his hand that day, had reached out to him in some way, but I didn't have the guts. Instead I rose from the couch, telling him I hoped he felt better soon, as he nodded, and I walked out of the room. I never had the gallantry or poise to tell my father how much he meant to me to his face. But I made a career of writing letters to him over the past few years, and he thanked me by sending me a postcard from New York Hospital, where he stayed for a few days after his second stroke. The postcard was a picture of the new wing and on the back he wrote something along the lines of:

> *Dear Al,*
>
> *Thanks for your kind letter. Do you like the scenery? It isn't the most exotic place I've ever been, but it's the only place I get to go anymore.*
> *Love,*
>
> *Your Dad*

We stopped our walks in late April. He was taking three or four naps a day by then and refusing to see anyone outside the immediate family. He spent most of the summer on the porch in Connecticut, always in a sweater and jacket, even on the warmest days. I drove to the Easton library and tried to pick out books he might like, but he grumbled over them all. One sunny July day, one year, three months, and twenty-four days before his death, he sat with a tray of food on his lap and a napkin tucked into the silk handkerchief around his neck.

"Hi Dad," I said.

"Hi."

"Looks good, is it celery soup?" I asked.

"How should I know, can't taste it," he said, eyes on the soup.

"Did you read any of the books I gave you?"

"They're all crap. Stop buying them. It's just a waste of money. Can you return them?" he said.

"I checked them out from the library."

"Well, stop checking them out. How much is the late fee?" he said.

"You're just like your mother. You like to throw away money. I don't need your help to find something to read. What do you know? Now get out of here. I'm taking a nap."

As time passed, I once again gave up on my father and our relationship. We barely talked for the next year, and then he died the morning after Thanksgiving of 2003.

My last conversation with him was on Thanksgiving night. I called him from a car in the middle of New Jersey on the way to dinner with a boyfriend's family. My mother told me he had stayed in bed all day. Perhaps I should have turned around at that news, but it didn't seem significant at the time. She put him on the phone:

"Hellooo," he hollered into the phone. His hearing was almost gone.

"Hi Dad," I said. "I'm sorry to hear you're not feeling well."

"I feel shitty all over," he said.

"Yeah, I know. I'm sorry. I'll be home tomorrow, but I wanted to wish you a Happy Thanksgiving."

"Happy Turkey Day to you, too."

And that was it. Not a particularly poignant or meaningful conversation, but it was ours, and did we really need a grand performance for the sake of the finale?

Trumpets are what I remember best from his memorial service. They were long and brass and moved with the same elegance my father once did. The two men who played them stood in the corner of the church like identical twins at a high school recital. They had gaunt frames and wore black suits, and their cheeks puttered like balloons teased with the promise of air.

They played Vivaldi's "Concerto for Two Trumpets" as we walked in from the side door of St. Bartholemew's Church, and I stared down the aisle at the crowds overflowing from pews, out the church door and down the stairs. There were faces I hadn't seen in years and years. People I didn't know were still alive, some hooked up to oxygen tanks.

We sang "Onward, Christian Soldiers," my father's favorite hymn. My mother spoke the lyrics as if she were talking to him. I didn't attempt to sing. I couldn't open my mouth. Instead, I pictured my father as a young man in his church pew on the Princeton campus singing the hymn, shoulder to shoulder with a fellow classmate, a future soldier.

Onward Christian soldiers,
marching as to war,
with the cross of Jesus,
Going on before!

I watched from the first row of the church as friends and grandsons spoke. Then more trumpets. I was told to rise. The cadence swelled, reaching its climax and momentarily washing the grief away. We proceeded out the side door, and my tallest nephew put his arm around me, and I was handed a baby wearing a ruffled French outfit, my father's last great-grandson, born only months earlier. I was kissed and hugged and introduced to wrinkled men in wheelchairs wearing finely tailored suits and soup-stained Hermes ties and handsome tortoiseshell spectacles, and then the trumpeting men ceased, and I watched as they packed up their instruments, looking both wan and exhausted like tuberculosis patients leaving for a clinic in France.

The baby in my arms tugged at my earring. I kissed his forehead pressing my chapped lips into his soft, sweet skin. His big, blue Weaver eyes stared up at me and I felt his heartbeat drum with life. Life. I got what he was trying tell me, and I squeezed him closer.

Through the falling snow, I could see my mother on the curb hassling the driver for not cleaning the cars before he picked us up. It was snowing; gray slush and pellets of salt covered the city streets, but she insisted the driver take the cars to the car wash before we pull up in front of the Colony Club, where the reception was being held.

The family huddled into each other on the street. Half-sisters all with distinct Weaver features, nephews in galoshes and macintoshes, great nieces in frilly dresses. My Swiss brother-in-law had his arm around my half-sister. He nodded as if to reassure me. I nodded back. The night my father died, this brother-in-law and my half-sister were in Switzerland.

Dad in costume for the last time

When they finally returned my mother's call, he asked to speak to me, his calming Swiss voice reverberated through the distant phone cords. He said he loved me. Weavers didn't tell each other they loved each other. I was touched.

Standing outside the church as the eight black limos drove east toward the car wash, I closed my eyes and saw my father standing in the far corner of our pool in Connecticut, where the cement ledge met one end of the green hedges. He had a hand on either side of the hot, gray blocks, and was wearing his blue-and-white-striped bathing trunks with the old-fashioned gold buckles that no one wore anymore; the sun falling in a perfect stripe across his slightly sagging but still athletic chest of sunspots and freckles.

"All the way with Uncle Jay!" he'd call to me as I bounced nervously on the third step of the shallow end, plugging my nose and diving through the water. I swam with all my might in hopes of making it to his waiting hands before I grew tired and began to sink, but his grasp was always there, even if I stopped midway.

34

CURRENTLY

My mother now lives alone in our long and winding Sutton Place apartment. It's been over two years since my father died, and his closet remains intact, his Turnbull & Asser shirts still meticulously folded and stacked in rows of ten, his pinstripe and tweed suits hanging one behind the other. In their shared bathroom in Connecticut, the mirrored shelf above his sink still holds his worn shaving brush and his comb, and his half-full prescription pill bottles line the top of the medicine cabinet.

I call her a few times a week to check in. I tell her to go to a widow's group, and she laughs and says those groups are for people with no friends. She pretends to be fine, but I detect a sadness in everything she does. Today she told me her doctor was worried, that her blood pressure had skyrocketed, and that she was forbidden to think about things that upset her. The doctor loaded her up with new prescriptions, upped her dose of Prinivil, and insisted on an EKG.

The first time I visited my father's grave I wept continuously for over an hour. It was February 15, two and a half months after he died. I lay like a baby across the frozen ground and kissed the cold stone cross. I traced the upside down triangles in the W of our last name. I spoke out loud to him, begging him to give me some sort of sign that he knew I loved him, that he knew I was okay. But there was nothing, just the cold February wind howling through the skeletal branches of the trees in Manchester, Vermont.

The second time I visited his grave I cried some more and arranged nine white tall-stemmed lilies at the base of the gigantic cross. It was nine months after he died.

But the last time I went up there, one year and three months after his death, I couldn't find his grave. The place was large, spread out over many rolling hills and wooden paths, and I must have wandered among the headstones for twenty or thirty minutes. This time I wasn't crying and I held a mixed bouquet of tulips and lilies. After nearly an hour, I found the enormous cross jutting out from the white earth, towering with greatness above all the others. Of course, my father had to have the biggest one; he was better than all the dead around him, better than the living above him. For a moment, I felt my grief dissipate and hatred take its place: hatred for my father's narcissism, for his egotism, for everything he refused to understand. I sat at the bottom of his headstone. A chubby-faced robin appeared on the branch of the tree my half-sister had planted next to the grave, his red breast shining in the bright winter sunlight. I sat there until the hatred left me and I kissed the stone W and left.

I haven't been back to his grave since.

It has been two years and four months since he died and sometimes when I am alone in my bedroom, I listen to the Vivaldi concerto they played at his funeral. There is something in the resonation of the chords of that piece, some strength that reminds me of my father. Each note sounds so complete, full of resolution and harmony and life, and when I play it I can feel him near me, breathing and laughing inside each note. And the man in the music isn't sad or angry. He's full of love and joy and he knows I am okay, and it is only inside the music we are able to love each other again.

There is no real way to deal with everything we lose in this world. The only way I've survived it is by holding onto memories, replaying them, temporarily moving into them when necessary. Somehow I seem to have held onto at least one thing from everyone I have loved and lost, even though many of these relationships ended in betrayal and heartbreak. I miss them all every day.

Last summer I spent a lot of time in the woods, cutting leafy needles and long, thorny plants from an overgrown path in Granville, New York,

a half hour from where my father lay in the ground. My bare feet sank
into the marshy turf below me, and horseflies and mosquitoes buzzed in
figure eights around my head. A tractor hummed in the distance and my
dogs darted ahead of me up the hill, intoxicated with life. The air was
thick with humidity, smelling of wet soil and wild flowers. Streamers of
incandescent pink shot across the pale blue sky as the tired sun fell behind the tree tops, its light, diluted by densely leaved branches, pattern-
ing the earth. I walked to the bottom of the path where the cool, fresh
pond water lapped the jagged shore. I dove in and swam in circles. And it
was only there in the water in the middle of these woods that I felt something close to whole. The truth of it is, even after all the ugliness, I
haven't entirely lost that thin-boned, shirtless child in muddy shorts wandering alone through the Connecticut woods. And that to me is the most
relieving thought of all.

ACKNOWLEDGMENTS

I'd like to thank Maya Rock, my agent at Writers House, for believing in this book from the beginning; my editor, Alison Stoltzfus, for her meticulous editing and for protecting my artistic integrity; Sydney Offit for inspiring this book four years ago in an undergraduate writing class; Honor Moore for her infinite wisdom and friendship; Susan Robertson for her patience and insights; Dr. Gregory for translating; and Tom Tarantino for his loyalty and attention to detail.

The crew: Melanie Greenberg, Jackie Nasser, Jay Kistler, Kate Garrick, David Goodwillie, Kristin McGonigle, Robert D'Aquila, Terra Becks, Ian Bentley, Stamatis Birsimijoglou, Elizabeth Koch, Lauren Grey, Travis Harvey, Percy and Ben, Julie and Benno Wurts for their infinite support and friendship. I would never have made it here without you.

Thanks also to the family: Wendy, Laurent, Julian, and Tristan Chaix; and my mother for surviving it all with me. I know much in this story saddens her and I applaud her for her bravery, for sticking by me through it all. I love you, Mom.